The Rural Landscape

Also by the Author

Southeastern Excursion Guidebook (1952), with Eugene Cotton Mather

The British Moorlands: A Problem in Land Utilization (1955)

The Southeastern United States (1967)

United States and Canada (1967)

Regions of the United States (1972), editor

The Look of the Land (1975)

Cultural Geography on Topographic Maps (1975), with Karl Raitz

The South (1976)

The Land That Feeds Us (1991)

Our Changing Cities (1991), editor

The Rural Landscape

JOHN FRASER HART

THE JOHNS HOPKINS UNIVERSITY PRESS
Baltimore & London

To Jean Blanding Fraser Hart
Who remained curious about everything and was always ready to go anywhere

Published in cooperation with the Center for American Places, Harrisonburg, Virginia

© 1998 The Johns Hopkins University Press
All rights reserved. Published 1998
Printed in the United States of America on acid-free paper
9 8 7 6 5 4 3 2 1

The Johns Hopkins University Press
2715 North Charles Street
Baltimore, Maryland 21218-4363
The Johns Hopkins Press Ltd., London

Library of Congress Cataloging-in-Publication Data are at the end of this book.
A catalog record for this book is available from the British Library.

ISBN 0-8018-5717-1

Contents

Preface and Acknowledgments

Sometime back around 1960, Merle Prunty told me that I reminded him of the man who jumped on his horse and rode off in all directions. "You have written about everything under the sun," he grumbled, "from fences and barns and manure piles to small towns and villages, and they have nothing in common." They did have something in common, if only that they had attracted my curiosity. When I analyzed my thought processes, I realized that I had been curious about them because they are all part of the visible landscape. I am intrigued by everything I see, and I am eager to understand the visible landscape better. Why and how did it get to be the way it is? Who are its creators and its custodians?

I tried to put my wide-ranging curiosity about the visible landscape into context by developing a lecture called "The Look of the Land." That lecture grew into a full-fledged course, and then into a book that bore the same name. That book was one of the unhappy victims of the decision by Congress to tax commercial publishers on their inventories rather than on their sales, and the publisher declared it out of print. In 1984, when George F. Thompson first broached the possibility of having the Johns Hopkins University Press reprint it, I told him that I had learned so much since I wrote it that I had to write a completely new book instead of reissuing the original work.

I have been singularly fortunate in having been privileged to sit at the feet of many fine teachers who have also become my good friends. I am grateful for all I have learned from Merle Prunty, Malcolm Proudfoot, Jimmy James, Cotton Mather, Wilbur Zelinsky, Wilma Fairchild, Glenn Trewartha, Dick Hartshorne, Peirce Lewis, Neil Salisbury, Wayne Kiefer, Ennis Chestang, Ev Smith, Carl Sauer, Bob Platt, Estyn Evans, Sally Myers, John Hudson, Karl Raitz, Phil Gersmehl, Tom Hubka, Betsy Pyle, John Morgan, Bob Ensminger, Kathy Klink, Wayne Price, and Chris Mayda. The most basic lesson I have learned from all of them is to keep my eyes, ears, and notebook open, and my mouth closed, because I have never learned much when my mouth was open.

This new book has benefited enormously from the constructive editorial suggestions of John Hudson, Peirce Lewis, and George F. Thompson. I have taxed the word-processing skills, patience, and resilience of Jodi Larson and Margaret Rasmussen, and I am grateful for the superb cartographic efforts of Carol Gersmehl, Alan Willis, Mark Lindberg, and especially Mui Le. I appreciate the research leaves I have received from the University of Minnesota. Jane Lincoln Taylor has done a splendid job of copyediting, and she has been fun to work with. And I am often reminded of T. W. Freeman's dictum that the wives of authors are strong candidates for stained-glass windows.

The Rural Landscape

1   *Understanding Landscape*

The only proper way to learn about and understand the landscape is to live in it, look at it, think about it, explore it, ask questions about it, contemplate it, and speculate about it. Merely reading about it is a sorry substitute indeed, but this book provides some initial clues about the kinds of things to look for and to ponder. It emphasizes things anyone can see, the vernacular, common, ordinary, everyday things of the people who live on the land, because these things make the landscape what it is.

We need to get out and look and see things for ourselves, because the primary source for studies of the landscape, oddly enough, is the landscape itself. I do not know how one learns to become a careful observer, so I am unable to teach the skill, but I know it is essential. We also need to learn to think analytically about what we have seen, and to identify the features of the landscape in terms that make sense to, and can be used by, others. We need to learn to talk to and listen to and learn from the plain, ordinary people who are the creators, the inhabitants, and the custodians of the landscape. We must know enough to ask them reasonably intelligent questions, and even more important, to listen carefully to what they tell us. We are not going to learn much while our mouths are open.

We can understand the aspirations, the needs, and the values of ordinary people only by listening to them, because they put pen to paper rarely, reluctantly, and with much painful effort.

The student of landscape should try to see everything, but most landscapes are so complex and so variegated that they are virtually incomprehensible without careful analysis. The person who tries to look at everything may wind up seeing nothing at all. We must be selective. At any given time we must concentrate on a few carefully chosen features, or types of feature, but we must never allow ourselves to forget that the features on which we concentrate are related in various ways, some close and some not so close, to all the other features of the landscape. From time to time we must also consciously change our focus and concentrate on another set of features in order to get a feeling for the complete landscape. With time and experience we should be able to develop a kind of sixth sense that will alert us to interesting aspects of features other than those on which we are concentrating at the moment.

For example, in 1993 John Morgan and I took another look at the fences along the route Eugene Mather and I had traveled forty years earlier. We found that the fences had not changed much, but our sixth sense told us that a new highwayside residential pattern had developed along rural roads in the South, especially in the region we have called Spersopolis.

THREE PRINCIPAL COMPONENTS

The features of the landscape are so complex that it makes sense to group them into three broad categories—mineral, vegetable, and animal (human)—to facilitate their analysis. The three principal components of any landscape are: (1) the landforms, or the features of the land surface; (2) the vegetation, or the plants that cover the surface; and (3) the structures people have added. The human structures may be further subdivided into four categories: (3a) systems of land division; (3b) structures associated with the economy; (3c) house types; and (3d) agglomerations of houses into villages, towns, and cities, with all their associated features.

The aspect of the landscape that makes the most compelling first impression is the form of the land surface: plains, plateaus, hills, mountains. The student of the landscape must have a good command of the basic principles of geomorphology to comprehend the origin and character of the major and minor forms of the earth's surface. The landforms of an area usually influence its patterns of vegetation and the ways in which people can use it. The rocks that lie beneath the surface also provide the parent materials from which its soils are formed, and they may contain minerals that are worth mining.

The second principal component of the landscape is the plants that cloak the surface of the earth. Trees, shrubs, grasses, and other plants make significant contributions to the landscape. They are also essential intermediaries between the mineral and animal worlds, because they convert the minerals of the earth's crust into food that animals (including the human variety) can eat. The plants that grow naturally in an area are excellent indicators of its potential for human use, and the existing vegetation often tells how people have used and abused the land.

The third principal component of the landscape is the structures that people have added to it. Every human structure, even a castle in the sand or a smoke ring in the air, is born to serve some human need, and the form of the structure usually reflects its function or the need it was originally designed to satisfy. Function and form are more intimately related in some structures than in others. They are most closely related in simple, ordinary, workaday structures, such as barns, sheds, garages, and privies, that are necessary for daily living; they are most widely divorced in expensive, pretentious, ornamental structures, such as showplace houses. That is why the study of house types is so extraordinarily complex, difficult, and often unsatisfactory.

THE IMPORTANCE OF MAKING A LIVING

Making a living is the single most important human activity. It consumes the greatest portion of our time and energy, it generates the greatest number of changes, and it is more closely related to more other variables than any other human activity. Economic activities are so important, in fact, that some economists dismiss as irrational any activity that is not obviously motivated by economic considerations. Such thinking clearly is muddleheaded, because many human activities, even in economically advanced societies, are completely rational, although they may be nowise rational economically.

As a general rule, people cannot live where they cannot grow food. Many exceptions immediately come to mind: fur-trading posts, sawmills, and pulp and paper mills in the vast wilderness of the boreal forest; small fishing ports nestled in snug harbors along stern and rockbound coasts; and mining camps in bleak, forbidding areas where food, water, workers, fuel, and even building materials have been brought in from outside to develop rich mineral deposits. But some of these settlements are temporary, depending on a resource that is not renewable, and they will be abandoned when the resource is exhausted. And all of them are but a drop in the bucket when the total population of the earth is considered, for their significance is small indeed.

A much more significant exception is the city, where millions of people do not grow their own food, yet even this exception is more illusory than real, because no city could long survive without a reliable source of food, and for the majority, the source must be close at hand. The first cities of the Middle East, for example, and the great cities of western Europe grew up in the breadbaskets of their respective areas, and even today no sizable city lies beyond the outer limits of modern commercial agriculture. Furthermore, only a small part of the occupied surface of the earth is built up; for example, three-quarters of the people of the United States live in urban areas, yet these areas cover less than 3 percent of the total land surface of the nation.

Making a living by producing food (and fiber) also plays a major role in shaping the rural landscape. The systems of agriculture in a given area are an important factor in determining the number of people the area can comfortably support, and they probably influence the way in which the people are distributed within it. An agricultural system also influences the rural landscape by its effect on the land use and plant cover of the area, and by the kinds of human structures it requires.

THE IMPORTANCE OF CULTURE GROUPS

The term *culture* refers to the beliefs, values, patterns of behavior, and technical competencies that are learned and socially transmitted by groups of people. The beliefs and values of different culture groups have molded and modified the landscape. The "cultural baggage" of a group may influence the form and appearance of its structures, because its members may have their own distinctive ideas about how particular types of building should look. It may also show up in their predilection for a particular crop or a particular breed of livestock.

Some beliefs and values are held more tenaciously than others, and some have a greater effect on the landscape than others. The members of the group may be bound together by language, ancestry, religion, skin color, or other common ties and interests, alone or in combination. Snowmobilers, skiers, and "suitcase farmers" are just as much culture groups as are the Scotch-Irish or the Old Order Amish, and perhaps no culture group has had a greater impact on the American landscape than the middle-class suburbanites who practice lawn worship.

One of the most striking culture groups in the United States is the Amish, some 150,000 strong, who live mainly in Pennsylvania, Ohio, Indiana, and Iowa. The Amish have received far more attention than their mere numbers might seem to warrant, because most of us are so wrapped up in sham and pretension that we admire and are fascinated by those who live simple, straightforward lives. The Amish people still cling to traditional styles of dress, and they farm in the old-fashioned way, with horses rather than tractors. Road signs in Amish areas warn unwary motorists that many of the people still travel by horse and buggy, and evidence of horses is abundant beside the hitching rail at the courthouse square in the county seat (fig. 1.1). The Amish people have so few surnames that we can identify Amish areas quite easily in plat books simply by shading in the areas owned by persons having no more than a dozen surnames and their variants (fig. 1.2).

Farmsteads with two houses rather than one are a distinctive feature of Amish colonies, because the men retire at a relatively young age ("worn out by the hard labor of farming with horses," say their non-Amish neighbors). The parents turn over their farm to the younger generation and move into the "grandpa house," which may be either an addition to the main house or a completely separate structure.

Many Amish farmsteads have Pennsylvania German barns. On one side is a barn bank, an inclined driveway that leads up to a central threshing floor on the second level. The second level projects on the opposite side of the barn as a forebay, or overhang, which is cantilevered out three to six feet over the stockyard. Many of the Amish barns in Iowa lack forebays, but the necessity of a forebay on a barn is so deeply ingrained in the thinking of the Amish farmer that each such barn has, in lieu of a forebay, a pent roof or "overshoot" that runs the length of the barn and projects out over the stockyard just as a regular forebay would have done.

The traditional Amish culture area in Lancaster County, Pennsylvania, is under enormous pressure because it is so close to the great cities of Megalopolis. New residential areas, shopping centers, and industrial parks are commandeering former farmland, and three million tourists clog the area each year. The Amish have high birthrates and large families, and agricultural land has become so scarce and so expensive that fathers can no longer afford to set up each of their children on their own farms. Amish young people have had to migrate to new Amish colonies in distant areas where land is cheaper, or they have had to find local nonagricultural jobs, and greater exposure to the non-Amish world can seriously threaten their traditional agrarian beliefs and values.

The beliefs and the values of the people who own, inhabit, and tend

Fig. 1.1.   Horse-and-buggy signs warn motorists that they are entering Amish culture areas.

the land will influence the form and appearance of the structures they need, and it is all too easy to misunderstand those beliefs and values. An Irish landscape-preservationist probably overstated his case, but he spoke for many others when he complained that the countryside is at the mercy of the farmers who own and develop it. He seemed to have forgotten that these very farmers and their forebears created the countryside he wants to preserve, and their livelihood depends on the skill and care with which they use and maintain it.

OTHER INFLUENCES

The human structures on the land are the product of a host of individual decisions made by a host of ordinary human beings. The imperative of making a living is the principal factor that influences most of these decisions, and that is why I prefer to invoke function to explain as much of the built landscape as possible, but the cultural baggage of the people, especially their beliefs and values, will also affect the form and appearance of the human structures in any area. These structures will also be influenced in varying degree by many other factors, including the nature of the physical environment, the technical competence of the builders, the date of construction, aesthetic and symbolic considerations, and the whims and idiosyncrasies of individuals.

The Physical Environment

The simple vernacular structures of ordinary folk usually are built of whatever material comes most readily to hand, and thus they often reflect the physical environment because it provides the stuff that people have to work with. Take the humble fence. In many parts of lowland England a patchwork of fields and pastures is stitched together by hedgerows of hawthorn, but in upland areas, where hawthorn does not do well, massive walls of stone faithfully reflect the underlying geology (fig. 1.3). In recent years, as labor has become more expensive, farmers have patched or replaced stone walls and hedgerows with barbed wire, and they have used barbed wire for most new enclosures, to the great dismay of hedgerow-lovers.

In the wooded areas of eastern North America most of the early settlers put up zigzag fences of rails split from the trees they had to clear from the land before they could cultivate it (fig. 1.4). They built stone walls in the stony, glaciated areas of New England and in areas where the limestone

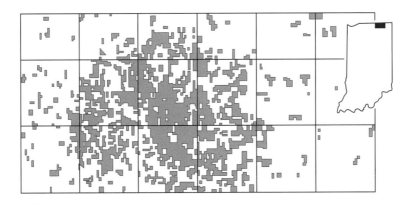

Fig. 1.2. The Amish area southeast of Goshen in Elkhart and Lagrange Counties, Indiana, is identified by shading the areas owned by persons named Miller, Yoder, Bontrager, Hostetler, Gingerich, Mast, Troyer, Schrock, Lambright, and Eash. Miller and Yoder alone account for 472 (51 percent) of the 920 properties.

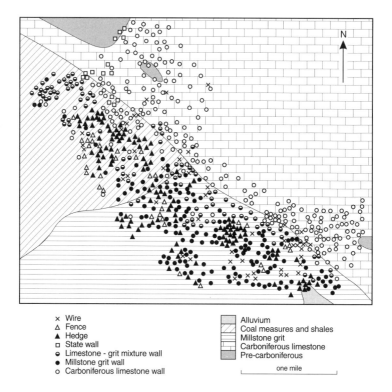

×	Wire
△	Fence
▲	Hedge
▫	State wall
◒	Limestone - grit mixture wall
●	Millstone grit wall
○	Carboniferous limestone wall

	Alluvium
	Coal measures and shales
	Millstone grit
	Carboniferous limestone
	Pre-carboniferous

one mile

Fig. 1.3. Field boundaries in the village of Ingleton (West Riding of Yorkshire, England) reflect the underlying geology. Based on a survey by Cedric March. Reproduced by permission from W. R. Mead, "The Study of Field Boundaries," *Geographische Zeitschrift* 54 (1966): 114, published by Franz Steiner Verlag, Stuttgart (formerly Wiesbaden).

bedrock formed easy natural flagstones. On the sandy outwash plains of the cutover lands of Michigan and Ontario, farmers uprooted the white-pine stumps the lumbermen had left and tipped them on their sides to make picturesque stump fences.

The treeless grasslands of the continental interior were not the deterrent to settlement that some people have supposed, but nonetheless settlers from wooded areas did have to scratch around a bit to find substitutes for wood for fuel, for buildings, and for fences. They experimented with walls of sod or mud and with hedgerows of Osage orange before the invention of barbed wire facilitated the spread of settlement westward across the grasslands. In the post-rock country of north-central Kansas the settlers even quarried square columns of stone and used them rather than wood for fence posts.

The use of barbed wire, made in distant factories and shipped to the grasslands at no small expense, shows how a prosperous economy can enable human structures to transcend the limitations of the local physical environment. Building materials for ordinary structures usually are not shipped far, but people in a prosperous area can afford to transport them from a greater distance than people in a poorer area. Steady improvements in the technology of transportation have enabled people to import building materials from ever greater distances, especially for public and ornamental buildings, and in contemporary North America few distinctive building materials are derived directly from the resources of the local environment.

Technical Competence

The technical competence of the people who build them, and the technical competence on which they can draw, also influence the form and appearance of human structures. Barn roofs illustrate the importance of technical skill. Almost any good carpenter can put up a straight roof, but the loft space under the eaves is cramped, and it requires much bending,

Fig. 1.4. Pioneer settlers built zigzag fences of rails split from the trees they had to clear from the land before they could cultivate it.

stooping, and head-bumping. Gambrel roofs, with gentle upper sections and steep lower sections, permit easier access to the space under the eaves, but building them demands greater skill; arched roofs, which are even more efficient, are also the most difficult to build. Most barns in the Midwest in the nineteenth century had straight roofs, but gambrel roofs gradually became more common as carpenters learned how to build them, and many barns built since the Depression have had arched roofs whose principal structural members have been bought prefabricated at the local lumberyard.

Date of Construction

Its date of construction also may affect the form and appearance of a building, both because improvements in technology over the years have given builders greater leeway and because the style of a building may reflect the changing fads and fancies of current fashion. House types are a good illustration. In the American colonies before the Revolutionary War, buildings were patterned after models in the mother country, and the austere elegance of the formal Georgian style was high fashion (fig. 1.5). After the Revolution the young republic turned for its model to ancient Greece, the cradle of democracy, and the Greek Revival style imitated the white pillars and open porticoes of Greek temples (fig. 1.6). The flamboyance of the Victorian era replaced the Greek Revival style in the latter part of the nineteenth century, and was replaced in turn by the California bungalow of the interwar years and the no-nonsense functionalism of the ranch style and split-level ramblers after World War II.

Aesthetics

Aesthetic considerations undoubtedly affect the form and appearance of many human structures, but it is easy to assign undue importance to them. Most people are motivated by functional, not aesthetic, considerations when they erect a structure, and most ordinary human structures must be understood in terms of their functions. They are not intended as works of art, and any artistic quality they may happen to possess is unconscious, accidental, and incidental. People do not intentionally erect structures that are ugly, but neither do they erect structures because they are beautiful—they erect them because they need them, not because such structures will beautify the landscape.

Fig. 1.5. The formal Georgian style of mansion was high style in the American colonies.

Most of us would like to live in a fine house, drive a fancy car, and wear the latest fashions, but nothing becomes outdated faster than that which is most up-to-date, and few of us can afford the luxury of constantly changing styles. The way we actually live is a compromise between our desires and our income. We like beauty—even though our perceptions of what is beautiful may not always agree with the perceptions of others—but we are compelled to balance our desire for beauty against the dictates of our pocketbooks. Each of us harbors notions of rightness and propriety about the way things ought to be, and we try to implement our ideals, but we cannot always afford to do so. We do the best we can with what we have, and the results sometimes are quite handsome, sometimes less so.

Symbolism and Meaning

A search for symbolism and meaning in the form and appearance of the human structures on the landscape is even more debatable than a search for aesthetics. There is no question that structures can be symbolic, but for whom are they symbolic? Is their symbolism the same for everyone? Symbolism may be the fabrication of the outsider. The insider accepts

Fig. 1.6. The Greek Revival style was popular in the early days of the American republic.

structures as functional necessities, and outsiders who find symbolism in them may be revealing more about themselves than about the structures. A structure may indeed have symbolism for the insider, but that symbolism is rarely articulated. Perhaps it is unwise to start probing too deeply for symbolism that is subconscious at best.

When I ask farmers about their buildings, if they give me answers that go beyond mere plausibility and make real sense, I see no need to quiz them further. If I do go further, in fact, I run the serious risk of imposing my own ideas, opinions, values, and beliefs on them. Some people seem to need symbolism, however, and they have refined to a quite remarkable degree the knack of finding it where others do not even know it exists. Some of these people claim they are able to detect symbols in the landscape so subtle that only they can see them, which makes one wonder whether such symbols are actually even there.

Perceptions

My strictures about searching for symbolism and meaning in the landscape apply with even greater force to attempts to assign psychological power to it. No one could question the ability of the landscape to influence human attitudes, ideas, and behavior, but fashions in the perception and appreciation of the landscape have changed greatly through time, and the psychological impact of the landscape varies enormously from person to person. Yi-Fu Tuan, for example, complains about the boring monotony of endless fields of corn and soybeans on the flat plains of the Grand Prairie of east central Illinois, but I find myself positively exhilarated by the vast open vistas of one of the finest farming areas on the face of the earth, and I feel uplifted and rejuvenated each time I travel across it. Who is to say either of us is wrong?

There is a danger of elevating our opinions and prejudices to the status of universal truths. All of us assume that we are perfectly normal human beings, and it is far too easy for us to move from that assumption to the belief that all other normal human beings think and behave and react just as we do. In our heads, we know they do not, but in our hearts we feel they should, and we become irritated when others are irrational, that is, when they do not think or do as we would. The world would have been spared many headaches if planners and economic developers could have been made to realize that their values are not universal.

Different people perceive and assess the same landscape quite differently, and praiseworthy attempts have been made to measure differences in landscape assessment as objectively as possible, but the principal conclusion that seems to emerge from such attempts is the complexity of the problem. We human beings are a cantankerous lot, and we seem to become increasingly cantankerous as we get further from the constraints of having to make a living, as we have greater freedom to indulge our idiosyncrasies.

Exceptions

It is far too easy to be distracted by exceptions and to take the norm for granted. Every area has one-of-a-kind structures that can only be explained as the products of individual whims. For example, highly unusual structures such as round barns may be so intriguing that they divert attention from the standard, ordinary, workaday structures that make up the vernacular landscape. Students of the landscape must learn to focus on the norms; they cannot allow themselves to be seduced by anomalies, no matter how fascinating these might be.

RELICT STRUCTURES

The landscape is rarely stable, because people's needs and opportunities are constantly changing. Change may be slow, but it is inevitable. New structures replace those that are outmoded, but the process is piecemeal, and older structures stand cheek by jowl with those that have superseded them. Rarely indeed is the slate wiped clean for a completely fresh start. Every landscape contains elements of persistence and elements of change, a mixture of the old and the new. People cling to what is old and familiar and comfortable even while they are experimenting with new ideas, and the earlier structures of an area influence all subsequent structures. One of the fundamental principles of landscape study is *prior potior*, which means "the earlier is the more powerful," or "the first one influences all that follow." The landforms, the plant life, and the human structures of an area all bear witness to the past as well as to the present.

Relict structures on the landscape are intriguing mementos of times gone by. In the Corn Belt, for instance, which stretches from eastern Nebraska to central Ohio, many farmers have shifted from mixed crop-and-livestock farming to cash-grain farming since World War II, and they no longer keep any livestock, which would require fences. These farmers have permitted their fences to fall into sad disrepair, or have even removed them completely, and fences are rapidly disappearing from the rural landscape.

Back in the days when cotton was king, farmers in the South mounded up thousands of miles of low earthen terraces along the contour lines to protect the bare soil from erosion by slowing down the runoff from torrential summer rainstorms. These terraces became relict features when farmers stopped growing cotton, but still they snake through pastures and even through pine forests that have grown up on land that once was used to grow cotton (fig. 1.7).

Some relict structures have been recycled by conversion to new and different uses. In many a small town, for example, the old bank building on the principal corner is now a drugstore or tavern, and many of the stores along Main Street now sell goods far different from those whose names are carved upon their lintels. The old schoolhouse, built solidly enough to keep the students from trashing it, now provides inexpensive floor space for a taxidermy company (fig. 1.8).

Out in the country, many buildings also have been put to uses other than those for which they were originally intended. City folk can buy handsome coffee-table books telling them how to convert old barns into

Fig. 1.7. The earthen terraces in the pasture in the foreground indicate that it once was cultivated for cotton. Similar terraces snake through the woods in the distance.

distinctive modern residences, and the home and garden sections of metropolitan newspapers regularly carry stories about old farm buildings converted to residential use. At a more rudimentary level, the farm family that gets running water for the first time may convert the old privy to a toolshed, move it from the back to the side yard, and give it a fresh coat of paint to advertise that it is no longer needed for its erstwhile function.

MANY KINDS OF INTERPRETATION

The form and appearance of human structures on the landscape are influenced by so many different considerations, from personal eccentricities to the necessities of making a living, that they have attracted the interest of scholars and artists with many different perspectives. Scholars and artists take different approaches to the study of the landscape. They begin with different basic assumptions, they emphasize different aspects of structures, they collect different kinds of evidence, and they lean toward different interpretations and explanations. The approach of artists is affective and the approach of scholars is analytical. There are no hard-and-fast lines

Fig. 1.8.   The old schoolhouse, which was built solidly enough to keep the students from trashing it, provides space for a taxidermy company.

between these two approaches; they complement and enrich each other. Scholars and artists each make an important contribution, and they need to work together, in a spirit of mutual respect, toward their common goal of a fuller understanding and appreciation of the landscape.

Contemplation of the landscape arouses feelings that range from subconscious to strong, and we can appreciate the landscape more fully when we understand why it affects us as it does. The approach here, however, is analytical. It begins by identifying the three principal components of the landscape, and then it explores each one in turn.

For example, the first step in an analysis of the structures that people have added to the landscape is careful examination and precise measurement of individual structures. Such work emphasizes the technical details of construction, and it leads to schemes of classification. Classification is essential, but it can easily become an end in itself, and some quite elaborate classification schemes are little more than busywork. A classification scheme is useful only to the extent that its categories are related to other variables of interest. The development of a tidy filing system is not a rewarding intellectual exercise if it allows us to perform only the clerical task of information retrieval.

A good scheme for classifying the human structures on the landscape is a prerequisite for extensive surveys of large areas, because we need to become so familiar with types of structures that we can recognize and identify them on sight when we see them in the field. A considerable apprenticeship is necessary to develop this familiarity, but it can snowball once it has been mastered. Extensive surveys of large areas can easily degenerate into superficiality in the hands of a novice, but they can be enormously revealing if they are conducted by a competent scholar.

Extensive surveys are concerned with the relationships of structures to their environments, and to the needs and values of the people who built them, especially the ordinary folk who leave such scanty written records of their tribulations and their triumphs, of their hopes and their dreams and their aspirations. Such surveys concentrate on the external appearance of structures. They may be subject to error, because the same facade can mask two quite different structures, especially if their interiors have been greatly remodeled, but surveys are necessary to put structures in their larger contexts.

The detailed examination of any individual structure requires a considerable amount of time, and a single scholar cannot hope to study more than a limited number. The "ants" who measure floor plans sometimes envy the "grasshoppers" who conduct extensive surveys, which they unfairly denigrate as "windshield surveys." The two types of work are equally essential and mutually supportive. Surveys would not be possible if detailed studies had not produced the necessary vocabulary, but we could never hope to survey the structures in an area of any appreciable size if we had to examine each and every structure in great detail.

Part One

ROCKS

The most immediately obvious aspect of the landscape is the nature of the land surface, whether it be the snowcapped volcanic cones of the Pacific Northwest, barren mountain ranges shimmering in the heat of the California desert, the stark cliff and canyon lands of northern Arizona, the majestic peaks of the Rocky Mountains, the interminable flatness of the Great Plains, the rolling glacial plains of the Middle West, the glacially rounded hills of New England, the tangled mountain fastnesses of Appalachia, or the stream-etched plains of the South.

The shape of the land surface is influenced by the nature of the rocks that lie beneath it. These rocks are also the parent material from which its soils are formed, and they may contain minerals that are worth mining. A geological classification of rocks often is not especially useful to a student of the landscape, because geologists are primarily interested in the age of the rocks. The student of the landscape is more interested in the nature and character of the rocks, or their lithology, and their ability to resist erosion. A geologist, for example, might classify a rock formation as Ordovician or Silurian, but a student of the landscape would ask whether it was sandstone or limestone.

Geomorphology is the science that seeks to explain the features of the earth's surface. Geomorphologists are concerned with geological structure and geological processes. Structure refers to the nature and attitude of the underlying rocks: whether they be igneous, metamorphic, or sedimentary; whether they have been uplifted, fractured, or both; and if they are sedimentary, whether their original horizontal layers have been tilted or crumpled. Process refers to the transformation of the land surface by weathering and by the agents of erosion, which include running water, groundwater, wind, and moving ice (glaciers). The underlying geological structure influences the ability of these agents of erosion to work their will, because some rocks resist erosion far more effectively than others. An agent of erosion becomes an agent of deposition when it dumps part or all of its load of sediment because its load exceeds its ability to transport that load.

At base level a stream has just enough slope to keep flowing, but not enough to enable it to erode its valley any deeper. It meanders from side to side and creates a level bottomland, or floodplain, on which it deposits loose silt and clay, or alluvium, when it floods. The highest parts of the floodplain, oddly enough, are the natural levees along the banks of the stream, where the water loses much of its velocity and deposits most of its load of sediment when it overflows its channel. The floodplain slopes ever so gently down and away from the levees into low-lying, poorly drained back swamps along its margins.

The ultimate base level of all streams is determined by the ocean, because a stream cannot cut its valley deeper than the body of water into which it flows, but most stream valleys have a series of temporary base levels. Some are upstream from bodies of water, such as lakes or larger rivers, into which the stream flows. Others are upstream from rapids or waterfalls

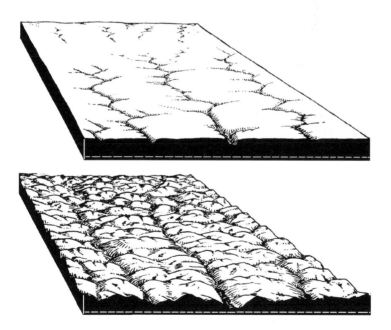

Fig. 2.2. Youthful topography (top diagram) has extensive level upland areas. Erosion by streams eventually produces maturely dissected topography (bottom diagram), with most of the land in slope. Most of the stream valleys in maturely dissected topography remain young. Reproduced by permission from Vernor C. Finch and Glenn T. Trewartha, *Physical Elements of Geography*, 3d ed. (New York: McGraw-Hill, 1949), 244.

Fig. 2.3. Maturely dissected topography in the Appalachian Plateau of southern West Virginia.

where the stream is having difficulty eroding an especially resistant rock formation.

Lakes and waterfalls are merely temporary geological features, and streams destroy them as expeditiously as possible. The outlet of a lake is a V-shaped valley that eventually cuts deep enough to drain the lake. Nature also clogs up lakes with aquatic plants whose remains settle to the bottom after the plants die, turning the lake into a peat bog. Streams destroy rapids and waterfalls by cutting V-shaped notches into them, and sometimes the water plunging over a waterfall undermines the rock beneath it and causes it to collapse. Waterfalls slowly move upstream. Niagara Falls, for example, is receding upstream at a rate of four to five feet a year. In fifty thousand years or so, Niagara Falls will have retreated all the way to Buffalo, the Niagara River will have started to drain Lake Erie, and people will have to find someplace else to go for their honeymoons.

A stream can start cutting its valley down to a new base level when it removes the lake or waterfall that caused the temporary base level. Flat remnants of its former bottomland may be perched as terraces on either side of the rejuvenated V-shaped stream valley.

TYPES OF ROCK

A stream drops all the material it has eroded from the land surface when it flows into the ocean. The coarser sediments are deposited closest to the land and the finer particles are carried farther out. Over thousands of years these sediments are gradually compressed and cemented into sedimentary rocks. The three principal types of sedimentary rock are classified by the size of the particles of which they are composed (fig. 2.4). Sandstone consists of sand particles that have been deposited fairly close to land. Shale consists of clay, silt, and mud particles that have been carried farther out to sea. Limestone is formed in deep water when calcium compounds that were dissolved in the water begin to separate out of solution and slowly settle to the bottom.

Coal, petroleum, and natural gas—the fossil fuels—are special types of sedimentary rock. They are formed from organic matter that accumulates in swampy areas where it cannot be attacked by the normal bacteria of decay. Coal is formed from plant tissues that originally decomposed into peat. The weight of the material subsequently deposited on the peat beds compresses the plant tissues first into lignite, or brown coal, then into bituminous coal, or soft coal, and eventually, under just the right circum-

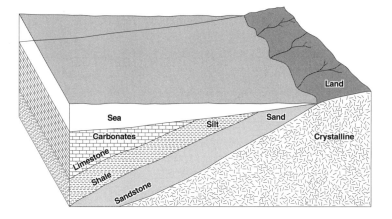

Fig. 2.4. Sandstone consists of coarse sand particles that were deposited near the shore, shale consists of finer silt and clay particles that were carried farther out to sea, and limestone is formed by the precipitation of carbonate minerals that were dissolved in the water. The conventional geological symbols are stippled dots for sandstone, horizontal dashes for shale, brickwork for limestone (see fig. 2.5), and randomly oriented dashes for crystalline rocks.

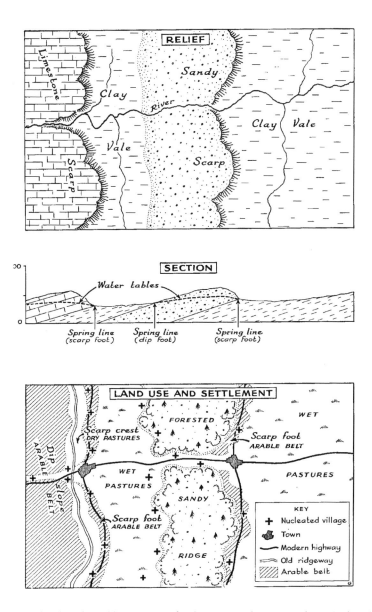

Fig. 2.5. The tilt and variable resistance of sedimentary rock strata, as shown in the middle diagram, influence the topography (top diagram) and land use (bottom diagram) of an area in southern England. The upland on the left, which is used to grow crops, is underlain by a resistant layer of limestone, whose edge forms an escarpment overlooking a clay vale. The clay vales are underlain by shale, which is so poorly drained that it can only be used for wet pasture. The resistant layer of sandstone forms an upland with an escarpment, but the sandy soil is so droughty that it is mostly wooded. Reproduced by permission from George W. Hoffman, ed., *A Geography of Europe* (New York: Ronald Press, 1953), 138. Copyright 1953 by Ronald Press. Reprinted by permission of John Wiley & Sons, Inc.

stances, into anthracite, or hard coal. Petroleum and natural gas are formed from animal matter rather than plant remains. They accumulate in porous reservoir rocks, such as sandstone. The reservoir rocks must be capped by impervious rocks such as shale to keep the oil and gas from seeping to the surface and escaping.

Sedimentary rocks originally form in horizontal layers, or strata, on the floor of the ocean. The strata may remain more or less horizontal when geologic forces raise them out of the sea and expose them to the forces of erosion, or they may be warped, arched up into domes, crumpled into folds, or broken along faults, either when they are lifted up or later. The dip, or tilt, and the resistance to erosion of the different strata affect the shape of the land surface and the varied ways in which people can use it (fig. 2.5).

Sandstone generally resists erosion in humid areas, but in dry areas it weathers easily. Soils derived from sandstone are porous and droughty, but they warm up quickly in the spring, and they are good for early crops. Shale is easily eroded, and soils derived from shale are cold, poorly drained, and sticky. Limestone generally resists erosion in dry lands, but in humid areas it slowly dissolves. Some limestone areas are honeycombed with caves and caverns (fig. 2.6). They have few surface streams, because rainwater drains through sinkholes on the surface and flows away in underground streams. Limestone areas generally have productive soils, but often these soils are thin, and they must be managed quite carefully to keep them from being washed away.

Sedimentary rocks generally do not resist erosion so well as crystalline rocks, which are either igneous or metamorphic. Igneous rocks form when molten rock material from deep within the earth rises toward the surface and begins to cool. This material may cool deep underground, but it may also reach the surface and pour out as a lava flow or erupt in a volcano. Metamorphic rocks are formed when other types of rock are transformed by intense heat and pressure. Crystalline rocks cannot contain deposits of fossil fuels, which would have been consumed by the heat to which these rocks have been subjected, but they may contain the ores of iron, gold, copper, nickel, silver, zinc, and other valuable metals. The quality of soils derived from crystalline rocks varies greatly, but in general these soils are not especially productive.

LANDFORMS IN ARID AREAS

Running water is the most important agent of erosion in humid areas, but arid lands, in which water is in short supply, have their own distinctive landforms. Surface features in arid areas are stark and sharp. Much bare rock is exposed. The alternation of blistering daytime heat and chilly nights shatters the rocks, and the lack of moisture limits the softening effects of chemical weathering. Few plants cloak the contours of the land. The wind blows constantly, unchecked by plants or by human structures. It carries away the finer particles of weathered material and leaves the surface littered with a desert pavement of stones too large for it to move.

Rain is rare in arid areas, but once or twice a year violent thunderstorms pummel the earth with great torrents of water. The water cannot soak into the sunbaked ground, and it is carried away by sudden flash floods. The floodwater is so choked with sediment that the streams cannot erode their valleys deeper, but they undercut the valley sides to create box canyons with steep sides and flat bottoms. The floodwater pours into shallow lakes and soon evaporates. The soils of arid areas contain large amounts of soluble

Fig. 2.6. In humid areas limestone is dissolved underground by percolating groundwater, and limestone areas are honeycombed with caves and caverns.

alkaline salts that would be removed by solution in a more humid area. Some of these salts are dissolved by the water that runs off in flash floods. When the water evaporates the salts are deposited on the dry lake beds as glistening salt flats caked with white alkaline salts.

Desert pavement, box canyons, and salt flats all are distinctive features of arid lands. Alluvial fans are not unique to arid lands, but they are sharper and clearer than in more humid areas. An alluvial fan is a gently sloping fan-shaped deposit that is formed at a sharp break in stream gradient, say where it flows from a mountain range onto a plain. The stream dumps its load of sediment and dams its own valley when its speed is reduced at the change in gradient. It repeatedly seeks and dams new channels, and gradually builds up a cone-shaped deposit of alluvium with its apex at the point where the stream leaves the mountains.

🌿 LANDFORMS

The rocks of an area reflect its geologic history, and they strongly influence, if they do not actually determine, the character of its surface features, or landforms. In 1914 Nevin M. Fenneman used geologic history and structure to divide the United States into major physiographic provinces. He assumed that the landforms in each province were similar, which was generally but not universally correct. In 1964 Edwin H. Hammond classified the landforms of the United States on the basis of their slope, relative relief, and profile. Relative relief, which is the difference in elevation between the highest and lowest points in an area, is probably the single most distinctive feature of landforms, because it is related to the degree to which they have been dissected by erosion. Plains have relative relief of less than three hundred feet, hills have relative relief of three hundred to one thousand feet, and mountains have relative relief of more than a thousand feet. Two distinctive landforms in the western United States are identified by their profiles: plateaus have deep canyons incised into level upland areas, and plains with scattered mountains are exactly what their name says they are.

The Ozark-Ouachita uplands of Missouri, Arkansas, and Oklahoma can be used to illustrate five different ways of showing the character of the land surface (fig. 2.7). Fenneman used the age and geologic structure of the rocks to identify physiographic provinces, Hammond used relative relief, and Erwin Raisz used small pictorial symbols. Richard Edes Harrison shaded the southern and eastern slopes darker and the northern and western slopes

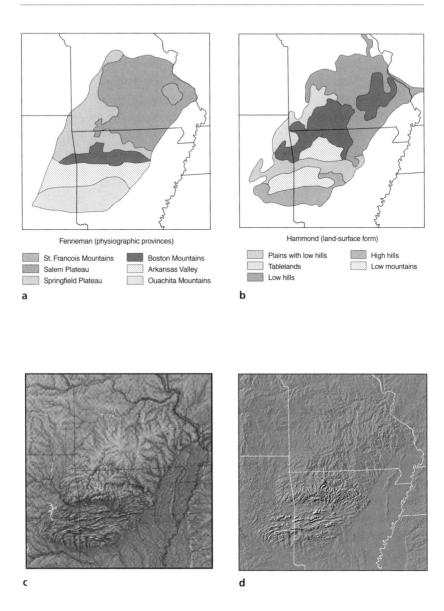

Fenneman (physiographic provinces)

St. Francois Mountains   Boston Mountains
Salem Plateau   Arkansas Valley
Springfield Plateau   Ouachita Mountains

a

Hammond (land-surface form)

Plains with low hills   High hills
Tablelands   Low mountains
Low hills

b

c

d

Fig. 2.7. The Ozark-Ouachita uplands of Missouri, Arkansas, and Oklahoma can be used to illustrate five different ways of showing the form of the land surface of a single area. (a) Nevin M. Fenneman used the age and geological structure of the rocks to identify six physiographic provinces (*The National Atlas of the United States of America* [Washington, D.C.: U.S. Geological Survey, 1970], 60). (b) Edwin H. Hammond used relief and profile to identify five categories of land-surface form (*National Atlas*, 63). (c) Richard Edes Harrison drew a map of

e

Fig. 2.7. *(continued)*   shaded relief by imagining that the surface was illuminated from the northwest (*National Atlas*, 57). (d) Gail P. Thelin and Richard J. Pike used computer technology to achieve the effect of shaded relief (Gail P. Thelin and Richard J. Pike, *Landforms of the Conterminous United States: A Digital Shaded-Relief Portrayal,* 1:3,500,000 [Reston, Va.: U.S. Geological Survey, 1991]). (e) Erwin Raisz, the finest artist ever to grace American cartography, used small pictorial symbols on his classic map, *Landforms of the United States,* 1:5,000,000, copyright 1957 by Erwin Raisz and reprinted by permission of GEOPLUS, P.O. Box 27, Danvers, Mass. 01923.

lighter to give the impression of a model of the land that is illuminated from the northwest. Richard J. Pike and Gail P. Thelin used this same technique.

The Ozark Plateau is a gentle geological dome. The oldest rocks are in the St. Francois Mountains south of St. Louis, where erosion has stripped away the overlying sedimentary strata to expose tough old igneous rocks that form high hills. These rocks contained deposits of iron ore, lead, and zinc, which have been mined. The rocks of the Salem Plateau are older than the rocks of the Springfield Plateau. Streams have dissected both areas into low hills except in the tableland of the northwest, which has extensive level uplands between its stream valleys. The streams that flow southward from the plateau are closer to the Gulf of Mexico than those that flow northward. Their base level is lower, and they have been able to dissect the southern and eastern parts of the plateau more deeply than the northern and western parts. The Boston Mountains at the southern edge of the plateau are capped with its youngest rocks, resistant sandstones that have been dissected into low mountains.

The Ouachita Mountains are underlain by tightly folded sedimentary strata that have been eroded into high hills and low mountains. The resistant sandstone strata form long subparallel ridges that follow the lines of the folds. The narrow valleys between the ridges are underlain by less-resistant shale that gives rise to inferior soils. The Arkansas Valley is a geological sag that was not uplifted so much as the Ouachita Mountains were. It has similar surface features, but the shale lowlands are more extensive, and the sandstone ridges are lower and fewer.

THE WORK OF ICE

Areas that have been glaciated also have distinctive surface features. The landforms of the northern part of North America show the effects of erosion and deposition by glaciers. Fifty thousand years ago, vast sheets of glacial ice more than a mile thick covered most of the area north of the Missouri and Ohio Rivers and Long Island, and mountain areas in the West boasted smaller glaciers. The ice sheet formed over Canada when more snow fell each winter than melted during the following summer. The weight of the deep accumulations of snow gradually compressed the lower parts into ice, and eventually the ice became so thick that it began to ooze outward in great tongue-shaped lobes.

The glacial ice eroded by plucking and scouring. Meltwater trickled down into cracks in the underlying bedrock when the sun melted the ice. This water froze into icy fingers attached to the glacier, and they plucked solid fragments of rock out of the ground as the glacier moved on. The angular chunks of rock frozen in the bottom of the glacier enabled it to act like a great sheet of sandpaper. It scraped away the soil, and it scoured, smoothed, rounded, and polished the bedrock across which it passed.

The material eroded by the ice and frozen in it was deposited helter-skelter across the countryside when the ice finally melted (fig. 2.8). The glacier created an undulating till plain when it melted fairly rapidly over a large area. Streams wander all over the till plain because the irregular deposits of glacial debris block normal drainage patterns. The soils of till

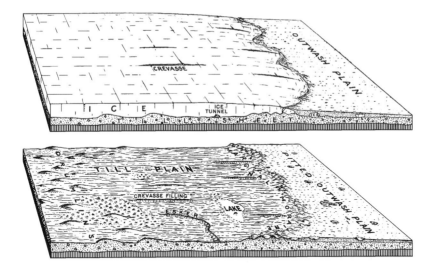

Fig. 2.8. The same area during and after glaciation. Undulating till plains were formed when the ice melted fairly rapidly over a large area, choppy moraines were formed where the edge of the ice sheet remained stationary for a period of time, and level outwash plains are underlain by the debris deposited by the sheets of water that poured off the melting ice. Reproduced by permission from Vernor C. Finch and Glenn T. Trewartha, *Physical Elements of Geography*, 3d ed. (New York: McGraw-Hill, 1949), 301.

plains usually are productive, but they must be drained artificially before they can be cultivated (fig. 2.9). Moraines are belts of choppier topography that were formed where the margin of the ice remained stationary for a period of time; the ice continued to move outward, but its margin remained stationary because the ice was melting as fast as it was moving outward. Moraines are hillier and stonier than till plains (fig. 2.10) because they received deposits of debris over a longer time, and usually they are wooded because their slopes are too steep to cultivate. Many moraines are studded with ice-block lakes. Blocks of ice were buried in the morainic debris, and when they eventually melted they left depressions that have filled with water and become lakes (fig. 2.11).

Beyond the moraine, at least in theory, is an outwash plain that was formed by the sheets of water that poured off the melting ice. This meltwater picked up some of the finer glacial debris and deposited it in level, crudely stratified aprons of sand and gravel beyond the edge of the ice.

Fig. 2.9.   The blotchy light and dark soils of till plains show the uneven deposition of glacial debris.

Fig. 2.10.   Some glacial moraines are so stony that they cannot be cultivated.

Fig. 2.11. Moraines are used for pasture or woodland when they are too choppy to be cultivated. Often they contain lakes in basins created when buried blocks of ice finally melt.

Most outwash plains are fairly sandy, and generally they are too droughty for successful cultivation, but they are ready-made sand and gravel pits, and near cities they are well-drained sites for large-scale residential development (fig. 2.12).

Other distinctive features associated with glaciation include glaciolacustrine plains and spillways (fig. 2.13). The Great Lakes drain northward through the St. Lawrence River, but this outlet was blocked during the glacial epoch by the ice sheets that were moving southward. The ice dams created large inland seas whose waters slowly backed up until they rose high enough to cut new temporary outlets southward. The great torrents of water flowing south cut deep trenches, glacial spillways, that are now occupied by the Minnesota, Wisconsin, Illinois, Wabash, Mohawk, and other rivers. The glacial spillways provided easy water-level routes into the interior of the continent before roads and railroads were built.

Fine sediments washed in from around their margins covered the floors of the glacial lakes. These glaciolacustrine plains were exposed as some of the flattest areas in the world when the ice dams melted and the lake waters could flow off to the north. These plains had to be drained before they could be cultivated, but areas such as the Red River Valley of

Fig. 2.12. The choppy wooded moraine at the top of the picture contrasts with the smooth outwash plain at the bottom. Outwash plains are good sites for large-scale residential developments, because they are flat and the sandy soil is well drained.

Minnesota, the Saginaw Plain of Michigan, and the Maumee Plain of Ohio were excellent agricultural lands once they had been drained.

Small glaciers have formed high on the flanks of many mountain ranges (fig. 2.14). These glaciers plucked away the rocks on their sidewalls and at their headwalls to create striking amphitheater-shaped bowls with nearly vertical rock walls. Small lakes of ice-cold, crystal-clear water often nestle in these bowls. When glaciers gouge out bowls on both sides of a mountain range they may reduce the headwalls between them to sharp, rocky, knife-edged divides that converge at spectacular pyramidal peaks. A glacier moving down a mountain valley will steepen its sides and widen its bottom, giving it a U-shaped cross-section. A moraine deposited at the lower end of the valley may dam up a long, narrow lake.

WATER BODIES ON THE LAND

Lakes are temporary geologic features, and natural forces destroy them as quickly as possible. When measured in geologic time even the Great

Lakes will be drained and disappear in the mere blink of an eye. Natural lakes are so rare, in fact, that they demand special explanation, and the explanation usually involves glaciers, limestone rocks, aridity, or volcanic activity.

Glaciated areas have three types of natural lake. Some are in rock basins that were scoured out by glacial erosion. Some are in moraines or outwash plains where large blocks of ice were buried beneath glacial debris. The subsequent melting of these ice blocks left depressions that have been filled with water. Some of the more picturesque lakes that are attributable to glacial action are long, narrow, moraine-dammed lakes in U-shaped valleys. The Finger Lakes of upstate New York are the best-known examples of such lakes, but many glaciated mountain areas have moraine-dammed lakes.

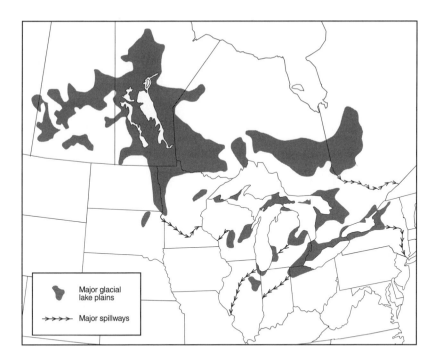

Major glacial lake plains

→→→→ Major spillways

Fig. 2.13. The great mass of glacial ice over Hudson Bay blocked normal drainage northward and created large inland seas that eventually overflowed southward through glacial spillways. The glaciolacustrine plains that were formed on the floors of these inland seas are some of the world's flattest areas. The glacial spillways were easy water-level routes into the interior of the continent before roads and railroads were built.

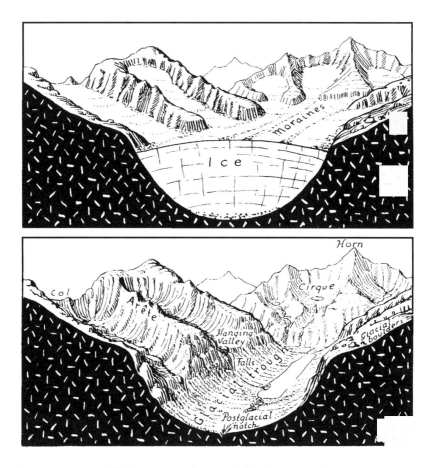

Fig. 2.14. An area of high mountains during and after glaciation. Glaciated mountain valleys have U-shaped cross sections. Reproduced by permission from Preston E. James, *A Geography of Man*, 3d ed. (Waltham, Mass.: Blaisdell, 1966), 509. Copyright 1966 by Blaisdell. Reprinted by permission of John Wiley & Sons, Inc.

Some areas underlain by limestone have large numbers of small lakes that might better be described as ponds. Limestone can be dissolved by groundwater (see fig. 2.6), and many limestone areas are drained through a honeycomb of underground passages rather than by surface streams. Rainwater enters these passages through sinkholes on the surface, and small ponds form in sinkholes that have become clogged with debris. They do not become large enough to overflow and create surface outlets, but they can disappear quite rapidly if the plug of debris is removed.

Ephemeral lakes are common in dry lands, where the rate of evaporation greatly exceeds the amount of rain that falls. The streams of dry lands flow into basins that have no outlets. After a rainfall the lowest part of a basin may have a broad, shallow lake, but the water quickly evaporates, leaving a glistening salt flat where the alkaline salts dissolved in the incoming water have been deposited. The Great Salt Lake in Utah is a well-known example of a dryland lake that once covered a much larger area.

Crater Lake in Oregon is one of the more unusual natural lakes in the United States. It occupies a depression into which an ancient volcanic cone collapsed because so much lava had been withdrawn from beneath it.

In most areas, bodies of water, if they even exist, are artificial impoundments. Various public and private agencies have created artificial lakes for many purposes by building dams. Many of the major rivers of the eastern United States have been turned into chains of lakes to improve navigation, to generate hydroelectric power, and to control floods. Many of these multipurpose lakes are also major recreational resources. Smaller streams in the East have been dammed to create reservoirs for municipal water supply, and mining operations often leave holes in the ground that become lakes or ponds when they are filled with rainwater.

In the western United States, huge dams have been built to create reservoirs that store the spring snowmelt of mountain areas for irrigation later in the growing season. Boulder Dam and the Grand Coulee Dam are the largest, but practically every stream flowing out of the Sierra Nevada in California has a dam and reservoir.

Many cattle-raising areas in the United States are speckled with small stock-watering ponds. The farmers or ranchers have created these ponds by damming up small streams to provide drinking water for their cattle, which need prodigious quantities; a mature cow or steer can drink a full bathtub completely dry in only two days.

AN INDEX OF MINING EMPLOYMENT

Geology determines the location of mineral deposits, but economics determines the location of mining areas. A geologic map shows where mineral deposits may occur, but many deposits are so small or so complex or so deep or so remote that they are not worth mining. We can identify mining areas by compiling maps from data on mining employment or on mineral production. The map compiler can simply shade each county according to the number of its workers employed in mining or the value of the minerals it produces, but a map that merely shows the total number of anything in each county is so misleading that it is virtually useless, because the areas of counties vary enormously.

Texas is 220 times the size of Rhode Island, and for most purposes it is ridiculous to compare the two states. The areas of counties vary even more. For example, consider three California counties in 1990 (table 2.1).

Table 2.1.

	San Bernardino	San Francisco	Sierra
Number of workers employed in mining	1,465	562	76
Area (square miles)	20,062	47	953
Mining workers per square mile	0.07	12.0	0.08
Total labor force	591,371	386,530	1,333
Mining workers as a percentage of the labor force	0.25	0.15	5.70
Index of mining employment	0	0	7

San Bernardino, the largest county in California, is 426 times the size of San Francisco, the smallest. The map compiler must standardize the data—that is, adjust them for variations in county size—by dividing the total number of mine workers in each county by the area of the county. In 1990 San Bernardino County had more than twice as many people employed in mining as San Francisco County, but San Francisco had many more mine workers per square mile.

The number of mine workers per square mile only tells part of the story, however, because many mining areas have few other sources of employment, and in some sparsely populated counties most of the people depend on mining for their livelihood. The map compiler must also calculate the percentage of the labor force in each county that is employed in mining to standardize the data for variations in the size of the population. San Francisco County had the greatest density of people employed in mining in California in 1990, but only one of every seven hundred workers in the county was employed in mining, whereas mining provided jobs for one of every twenty workers in Sierra County.

Two quite different maps result when the data on mining employment in California and Nevada in 1990 are standardized for variations in area and in population size. The greatest densities are near major metropolitan centers, but the highest percentages are in sparsely populated counties. It is apparent that some of the people employed in the mining industry, especially in metropolitan areas, are office workers who rarely if ever venture near a mine.

In the United States in 1990 one of every 160 workers, or 0.625 percent of the total labor force, was employed in the mining industry. I assumed that this percentage was the minimum basic requirement for a mining county, so for each county I subtracted 0.625 percent of the total labor force (-0.00625T) from the number of persons employed in mining (M) to calculate the number of "surplus" mine workers in the county. I divided the number of surplus mine workers by the area (A) of the county in hundreds of square miles to derive an index of mining employment (I) for the county (see fig. 3.1 below).

3   *Landscapes of Mining*

Mining areas are fascinating in a macabre sort of way. Mining is a dangerous and dirty business. The extraction of minerals often leaves ugly scars on the face of the land. The processing of minerals can produce great quantities of waste, which fouls the air with smoke and dust, and pollutes the waters with noxious chemicals. Many mineral deposits are in remote and isolated places, in environments so inhospitable that they could support only a sparse population were it not for mining: the icy wastelands of northern Alaska, the searing deserts of the Middle East and Africa, the precipitous chasms of the mountainous West, the steaming swamps and marshes of Texas and Louisiana, the rugged hills of West Virginia and Kentucky.

The mining company must develop roads, railroads, canals, and pipelines to bring in the workers and their tools and equipment, and to carry out the minerals once they have been torn from the ground. Workers flock in to seek their fortunes, and a sea of tents or shacks or trailers springs up almost overnight. The local people, if indeed there are any, are outnumbered and overwhelmed by the rough-and-ready newcomers, mostly husky, boisterous, single young men with few inhibitions. There are few jobs for women.

Mining camps, as their name implies, are merely temporary intrusions. A few mining areas have developed manufacturing activities, and they have been able to survive by importing their raw materials from other areas after their own mineral bases have been exhausted, but most mining camps are built to be abandoned. They are created solely to exploit a mineral deposit, and they have little use once that deposit has been mined out or is no longer economical to work. A drop in the price of the mineral it produces can turn a mining camp into a ghost town just as suddenly as it was created.

The principal mining areas in the United States in 1990 were mineralized areas in the mountain states, the "oil patch" of Texas and adjacent states, and the Eastern Interior and Appalachian coal fields (fig. 3.1). These same general areas also had the greatest concentrations of counties that produced more than five million dollars' worth of minerals in 1982 (fig. 3.2). The United States Census of Mineral Industries published data only for these counties, and for quite a few of them it suppressed much of the data under disclosure rules. The Bureau of the Census is explicitly forbidden to disclose any information about individuals or individual companies to anyone, and it scrupulously suppresses all data that might reveal such information.

The oil and gas business dwarfs all other mineral industries in the United States. It is the leading producer in more than half of the nation's mining counties (fig. 3.3). Coal mining is a poor second. The principal coal-mining areas are in the Appalachian field, the Eastern Interior field, and scattered parts of the West. The nonmetals, which rank third, include unexciting but essential building materials, such as crushed and broken stone and sand and gravel, which are produced close to the cities where they are used because they are too bulky to ship any great distance. The nonmetals also include dimension stone, such as granite and marble, and chemical and fertilizer minerals, such as sulfur and phosphate. Nonmetals may be quite important locally, but they are little more than a drop in the national bucket. The same is true of the ores of iron, copper, gold, silver, lead, zinc, and other metals.

OIL AND GAS

The distinctive landscapes of mining areas are produced by the extraction, processing, and transportation of minerals and by the ways in which the workers are housed. Oil and gas are extracted from wells. Often the only visible sign of an oil field is a scatter of squat black pumps that look like giant insects as their rocker arms bob rhythmically up and down. Unpaved access roads serve each pump, and buried pipes carry away the crude oil, whether to rusty clusters of field storage tanks or to refineries. In the old days many oil fields sported stands of gangling metal derricks that had been used to drill the wells, but nowadays workers use portable rigs that they can pack up and move to another site once they have completed a well.

Crude oil must be processed at a refinery, a gleaming maze of pipes and retorts that boil off and then condense its various chemical constit-

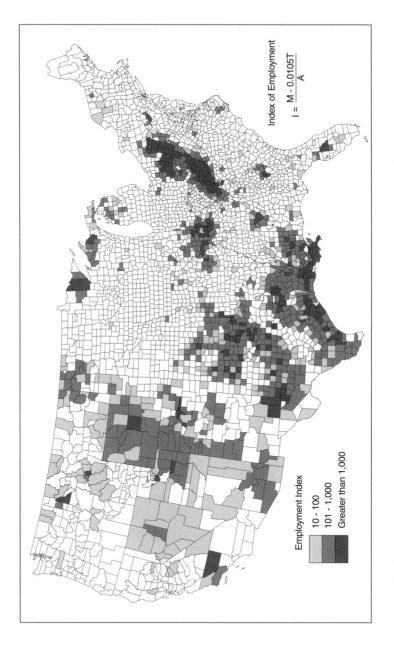

Fig. 3.1. An index (I) of mining employment in the counties of the United States in 1980. M is the number of persons employed in mining, T is the total number of persons employed in the county, and A is the area of the county in hundreds of square miles.

uents. Beside the refinery are rows of storage tanks of divers shapes and sizes to hold the different chemicals, and not too far away are petrochemical plants that use some of the products of the refinery to make basic industrial chemicals. Most oil refineries are near the fields where the oil is taken from the ground, but pipelines and tankers take some crude oil to refineries in the densely populated market areas where the products are used.

COAL

Coal, our second-most-valuable mineral, once was extracted mainly from underground shaft or drift mines (fig. 3.4). Drift mines are simply tunnels dug horizontally into a hillside, but most underground mines have vertical shafts with hoisting cables that raise and lower elevator cars. The headframe of a shaft mine supports large grooved wheels that guide the hoisting cables. Beside the headframe is the winding house. A revolving drum in the winding house plays out cable to lower the elevator cars with miners and machinery to the working level deep within the earth, and it rewinds to raise cars loaded with coal or with tired and dirty miners at the end of their working shift.

Near the headframe and the winding house are the lamp house, where the miners can change their clothes and shower, a nondescript clutter of repair shops and storage sheds, and the tipple, where the coal is sorted and prepared for loading onto the barges, railroad cars, or trucks that will haul it away to market. Transport by barge is cheapest, but many coal-mining areas are not near navigable rivers or canals, and they are laced with railroad tracks that carry long lines of open gondola cars loaded with coal.

At the working face underground the miners break loose the coal and load it into the mine cars that haul it back to the shaft to be raised to the surface. A certain amount of useless rock almost inevitably is taken to the surface along with the coal, and it must be removed before the coal is shipped off to market. This waste rock is dumped in huge, ugly tipheaps that are the unsightly landmarks of many coal-mining areas. Tipheaps may remain barren piles of broken rock for years, because the rocks often contain sulfur and other toxic chemicals that discourage plant growth.

The miners support the roof of the mine with massive wooden pit props after they have removed the coal, and in some mines they fill in the worked-out areas with waste rock brought back from the surface. Eventu-

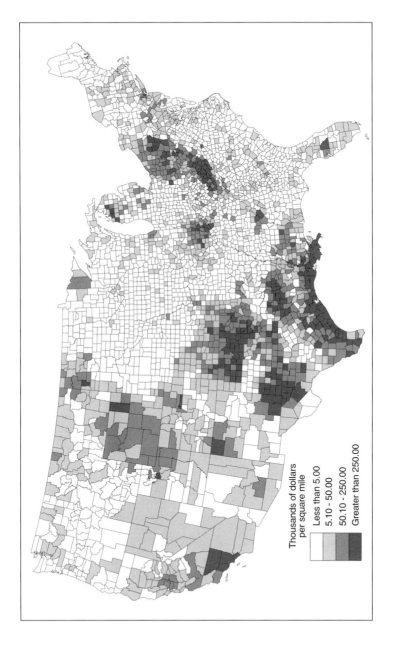

Thousands of dollars
per square mile

Less than 5.00
5.10 - 50.00
50.10 - 250.00
Greater than 250.00

Fig. 3.2. The value of mineral products in 1982 in counties that produced five million dollars' worth or more.

ally the pit props rot away, however, and the roof of the mine begins to subside. The uneven subsidence of old mines can wreak havoc at the surface, where houses sag and crack, roads buckle unexpectedly, water collects in strange places, and streams pond up and overflow their banks. The possibility of subsidence discourages the development of old mining areas because people are reluctant to invest their money in factories or office buildings that may start to sink into the ground.

STRIP MINES

In the early 1970s the principal method of mining coal in the United States shifted from underground mining to strip mining, which is practical only where a seam of coal thick enough is close enough to the surface (fig. 3.5). Strip-mine operators use some of the world's largest and most powerful earthmoving equipment. The steep steel stairs, the smell of grease and lubricating oil, the clang of metal, and the constantly throbbing machinery on a huge, lurching power shovel remind me of the engine room of a destroyer in a heavy sea.

A power shovel begins a strip mine by digging a trench, or box cut, down to the top of the coal seam (fig. 3.6). It dumps the useless overburden in a steep ridge, or spoil bank, along the far side of the trench. A smaller shovel digs out the coal at the bottom of the trench and loads it into trucks to be hauled away, leaving a barren ditch. The large shovel digs a new trench parallel to the first one, and dumps the spoil from this second trench into the barren trench from which the coal has been removed.

The excavation of subsequent trenches leaves long, roughly parallel ridges of jagged spoil banks that corrugate the mined-out area. The last trench remains empty, a deep narrow pit with a vertical highwall of solid rock on one side and a steep spoil bank on the other. In time, rainwater fills the last trench and the deeper depressions between the spoil banks. It turns them into long, narrow ponds, but the spoil often contains enough sulfur to poison the water for fish and to make it unsafe for swimmers. Your fingers feel slippery if you dip them in the water and rub them together, because the acid is eating away your flesh.

The spoil banks are merely piles of broken rock that have no topsoil, and rainwater converts their sulfur into sulfuric acid, which kills plants. Grading the spoil banks can restore the shape of the land, but the sour, acid

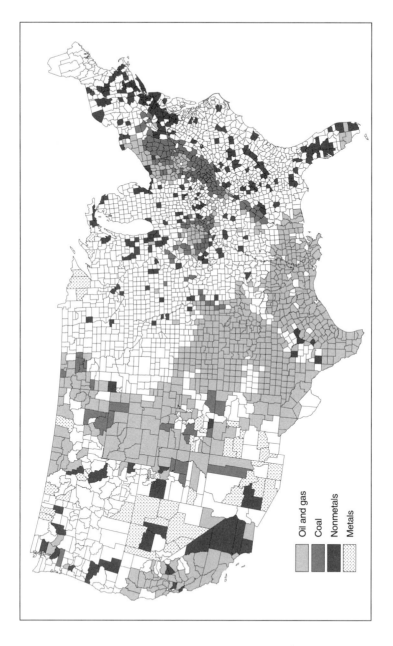

Fig. 3.3. The most valuable mineral product in each mining county in 1982.

Oil and gas

Coal

Nonmetals

Metals

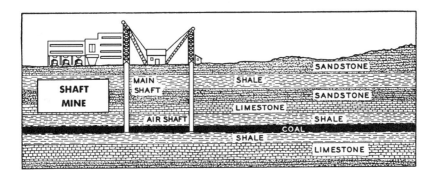

Fig. 3.4. A diagram of a shaft coal mine. Reproduced by permission of the Bituminous Coal Institute.

surface must be limed to reduce its acidity, and it must be fertilized to enable plants to grow on it. Ideally the original topsoil should be removed and saved before mining begins, and it should be spread back on the land after it has been graded, but most mine operators are understandably reluctant to pay for doing more than the law requires. An adequately reclaimed strip-mined area would be hard to distinguish from the undisturbed areas around it, but good examples are precious few.

Many states have laws to regulate strip mining, but the mine operators have had a vested interest in ensuring that these laws had no teeth. Most states have required the operators to post a bond that was not returned until they had reclaimed the land, but the bond has been so low that many operators have expected to forfeit it, and they have treated it merely as a nuisance tax on their operations. Action was required at the federal level, and in 1974 Congress passed a law to prevent the abuses of strip mining, but some administrators have been less than enthusiastic about enforcing it. Some states subsequently have tightened their laws requiring careful reclamation of stripped-out areas, and mine operators have been known to claim that they leave the land in better shape than they found it.

Strip mining does enormous damage on level land, but it is even more devastating in hilly areas. A power shovel can cut only a narrow bench into a steep hillside before the overburden becomes too thick, but the bench may snake for great distances along the contours, following the outcrop of the coal seam. Often the shovel simply dumps the spoil over the edge of the bench. The spoil goes thundering down the slope, knocking down trees and

Fig. 3.5. A diagram of a strip coal mine. Reproduced by permission of the Bituminous Coal Institute.

houses, burying fields and gardens, damming up creeks, blocking roads, and destroying everything in its way. The scars of strip-mined benches and their spoil screes high on the hillsides are eyesores that are visible for miles across the countryside.

PROCESSING COAL

Coal is an essential industrial raw material as well as a valuable fuel, and it may be processed in several different ways after it leaves the mine. The Industrial Revolution of the nineteenth century was not possible until coke replaced charcoal for smelting metallic ores. Coke is an almost-pure form of carbon that is made by baking coal to drive off its gases and tars. The first coke ovens were shaped like large beehives. Beehive ovens did not capture the gases and tars, which are important raw materials for the chemical industry, and they have been superseded by more-efficient but less-picturesque by-product ovens (fig. 3.7).

Coal is used as a fuel in steam engines and other machines, but its principal use as a fuel today is to heat water to produce steam to drive the turbines in coal-fired electricity-generating plants, which are marked by their slender, towering smokestacks. Most power plants are near bodies of water, both for cheap coal transport and for intake water to make steam. They used to discharge wastewater that was warmer than the water they took, which damaged aquatic life, and many of them have started to conserve water by condensing it in cooling towers. Huge, wasp-waisted concrete cooling towers have been a common sight in Europe for years, and they are becoming increasingly common in the United States.

NONMETALS

Nonmetals, the third-most-valuable type of mineral mined in the United States, are a heterogeneous group. Building materials such as crushed stone, sand and gravel, and cement are essential to a modern industrial society, but they are too bulky to ship any great distance. They are extracted as cheaply as possible and as close as possible to the place where they will be used. Most major American cities are served by local sand-and-gravel pits and by quarries where limestone is extracted for making cement. Many people are not even aware of these operations; often they are screened by plantings or by high board walls to conceal their unsightliness from the public eye. Areas that have the right kinds of rocks have such highly specialized and locally important mining operations as granite quarries in Vermont, quarries for monumental limestone in Indiana, kaolin pits in Georgia, open-pit phosphate mines in Florida, sulfur wells along the Gulf Coast, and

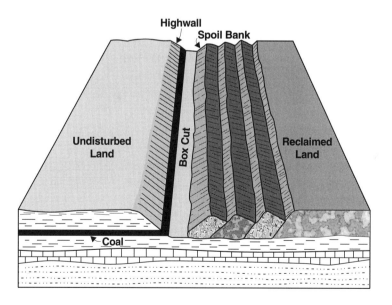

Fig. 3.6. A sketch of a strip coal mine with reclamation of the spoil banks already under way.

Fig. 3.7. Long, eerie rows of abandoned beehive coke ovens haunt the hillsides south of Pittsburgh around Connellsville, Pennsylvania. The front parts of the ovens have collapsed, and they look like gaunt, hollow sockets without eyeballs.

Fig. 3.8. An open-pit copper mine at Bingham Canyon, Utah.

open-pit phosphate mines in Florida, sulfur wells along the Gulf Coast, and open-pit mines for soluble minerals in arid areas in the West.

METALLIC ORES

Metallic ores are the least-valuable group of minerals mined in the United States, but they may also be the best known. The value of copper and iron ore mined in this country is only slightly greater than the value of sand and gravel, and the value of the gold and silver we produce is only a tenth as great as the value of our crushed and broken stone, yet many people probably think first of metals when they think of mining. Metallic ores are extracted in small but highly specialized mining districts that have produced great wealth in proportion to their size, and the mining of these ores has produced some spectacular landscapes.

Some metallic ores are taken from shaft mines, but metallic ores are mined from open pits whenever possible because even the best ores contain large amounts of unusable material that adds to the cost of handling. Even the richest iron ores are only about 50 percent iron, and the Homestake Mine at Lead, South Dakota, averages slightly less than three ounces of gold from every ten tons of ore. An unusual type of open-air mine was the placer mine in the early gold workings in California, which used water to sort the heavy nuggets and flecks of gold from large alluvial deposits of lighter sand and gravel. Miners did not begin prospecting for the mother lode from which this gold had been eroded until they had worked out the placer deposits.

Huge open-pit mines for extracting metallic ores are some of the largest landscape features human beings have ever created. The great pyramids of Egypt, for instance, could easily be lost in the copper mine at Bingham Canyon, Utah (fig. 3.8), which is half a mile deep and two miles wide, or the iron ore mine at Hibbing, Minnesota (fig. 3.9), which is three miles long and a mile wide. The miners scrape away the thin overburden, blast loose the exposed rock, and load it into railroad cars or huge trucks, which haul it to the processing plant, where as much of the waste rock as possible is removed to save the cost of shipping it to a smelter.

Most metallic ores must be processed both physically and chemically to remove their waste material. Workers process the ore physically at the mine. They crush it, which creates great quantities of dust, and they wash it to separate the heavier ore from the lighter waste. They pump the wash-

Fig. 3.9.   An open-pit iron ore mine at Hibbing, Minnesota.

water into a tailings pond where the waste can settle out. Many mine waste dumps and tailings ponds are unstable. From time to time one collapses in an avalanche of debris that can bury people alive and destroy large numbers of buildings.

Some tailings ponds are enormous. Before it went bankrupt, the Reserve Mining Company dammed several streams to create a tailings basin five miles west of its taconite processing plant in Silver Bay, Minnesota. The largest dam, 195 feet high and 2.5 miles long, was one of the ten largest earthen dams in the world. The basin covered 5.8 square miles and held seven hundred million tons of tailings. Rail cars hauled coarse tailings to the basin, and the finer tailings were pumped through pipelines in a water slurry.

Some of the waste material in metallic ores can be separated physically, but most ores also contain impurities that can only be removed by chemical processing at a smelter or refinery. Smelters and refineries require considerable capital investment, and they usually process the ore from several mines. They are more permanent than mines because they can continue to operate even when individual mines close down.

Some impurities are removed by dissolving them in acids or other strong chemicals. Others are removed by mixing the ore with another substance, such as coke, which will combine with the impurities when the mixture is heated until it melts. The impurities form a scum, or slag, that floats on top of the molten metal in the furnace. It is removed and dumped in unlovely slag heaps. Many smelters and refineries also produce noxious gases that kill plants and damage lungs. They have tall stacks to release these gases high in the air to dilute their effect. The big stack at the copper and nickel refinery in Sudbury, Ontario, is nearly a quarter of a mile high.

The epitome of a mineral processing area is the kind of heavy industrial complex that developed in the nineteenth century in major coal-mining areas such as the English Midlands, the German Ruhr, the area around Pittsburgh in the United States, and the Donets Basin in the Ukraine. Coal fired the by-product coke ovens that turned other coal into coke, and it fired the chemical plants where the gases and tars from the coke ovens were made into basic industrial chemicals. Coal fired the blast furnaces where iron ore was fused with coke to make pig iron. Iron ore moved to the coal fields, rather than vice versa, because so much more coal was needed to make each ton of pig iron. Coal fired the mills where pig iron was made into steel, and coal fired the factories where steel was made into boilers, locomotives, and other heavy machinery. The coal fields that have been developed in the twentieth century are more like other mining areas because they produce a mineral that is shipped out of the area instead of providing a basis for the development of manufacturing within it.

HOUSING

In most nonindustrialized mining areas, the distribution of workers' houses is related to the method of extraction. Miners can live almost anywhere in oil and gas fields, where the wells are scattered all over the countryside, or in strip-mining areas, where the power shovels are constantly rumbling to new locations, but miners' houses huddle together, often in company towns, near shaft mines and large open pits, where the mine entrance remains fixed.

A company developing a mine in an isolated area or in a difficult environment has to house the construction workers when they are sinking the shaft, removing the overburden, building the roads, and laying the rails, and then it has to house the miners once the mine is operating. The mining company might build its own company town, which would be dominated by the headframe, tipple, and other mine structures, and by a large building in the center that would hold the company store and offices.

The streets in company towns were laid out in tidy geometric patterns, but they were lined with long, grim rows of identical houses that were built as cheaply as possible. The foremen and supervisors lived in a small area of slightly better houses. The company could evict the workers if they lost their jobs or went on strike. It could simply abandon the houses when it closed down the mine, or it could try to sell them to the workers, who placed them on skids and dragged them down the road to another location. Nowadays company towns have been superseded by mobile-home parks, with units that can be towed away when the mines close down.

Company towns, like all mining camps, were temporary. They began life as boom towns, and they were doomed to become ghost towns, with devastated hillsides, abandoned workings, ugly mine dumps, polluted streams, derelict mine buildings, rusting machinery, long, dreary rows of tired and empty houses, battered and rutted roads, and grass growing in the railroad tracks. Ghost towns have a morbid fascination for many people, however, and some have been revived as tourist attractions.

Part Two

PLANTS

The second principal component of the rural landscape is the plant life that cloaks the surface of the land. The four basic categories of plant life are herbs, grasses, shrubs, and trees, which may be cultivated or noncultivated. The vegetation of any area contributes to the appearance of the present landscape, it can tell us much about the climate that environs it and the soil that nurtures it, and it can tell the tale of how people have used the area in the past and of how they might be able to use it in the future.

It may seem odd to begin a discussion of plant life by saying that animals cannot eat rocks, but this truism is one of the basic principles of life on our planet. Animals (including people) could not live if they did not have plants to convert the rocks and minerals of the earth's crust and the chemicals of the soil into vegetable matter they can digest.

Animals are free to wander, but they cannot wander too far from areas that are environmentally suitable for the plants they need to keep them alive, and plants are rooted in the soil; they cannot run, they cannot hide, and they are subject to every whim of nature. They must be able to withstand the bite of the wind, the scorching heat of the midday sun, the numbing cold of the long night watches. Their roots may be starved by a niggardly soil, alternately parched and waterlogged, or even laid bare by the ravages of erosion.

Weather and climate are the basic elements that determine the broad regional patterns of plant life, but these patterns are also influenced by other environmental variables, such as the shape of the land surface and the soil that covers it, and by the activities of people and their animals.

4 *Plant Life*

Plants obtain the food they need from the soil when their roots absorb moisture that contains dissolved chemical nutrients. They can only thrive in areas that have enough precipitation to maintain the necessary soil-moisture level, because they starve and wither away if the soil is too dry. The growth of plants is thus related to monthly and annual averages and variations of precipitation and temperature, which are the stuff of climatic classification. Most classifications of climate are based on the requirements of the more-important plant species, and that is why maps of climatic regions often resemble vegetation maps so closely.

Large-scale patterns of vegetation reflect climatic conditions, but within each climatic region the vegetation varies in relation to differences in elevation, slope, exposure, drainage, and other environmental factors. It is also influenced by climatic variables, including potential evapotranspiration (the loss of moisture by evaporation from the soil and by transpiration from plants), the length of the growing season, the frequency and severity of droughts and freezes, the amount and persistence of snow, and the highest and lowest temperatures at critical times in the life cycles of particular plants.

Seasonal weather patterns are decisive for the sensitive stages in the life cycle of a plant species, which include flower production, pollination, seed set, seed germination, and seedling establishment. For example, a spring frost that does not affect the other parts of a plant may kill its tender buds and flowers and keep it from producing seeds and bearing fruit that year.

At the local level, variations in topography and soil can greatly complicate the broad regional patterns of plant life that are related to weather and climate. For example, plants that thrive in sheltered valleys may not even be able to survive on windswept ridges. Slopes that face south toward the sun usually are warmer and drier than shady north-facing slopes (fig. 4.1). Excessive soil moisture that accumulates in poorly drained low-lying

Fig. 4.1. The south-facing slope on the left, which is grass covered, is warmer and drier than the north-facing slope on the right, which is tree covered.

areas may kill plants by suffocating their roots. Low-lying areas may also be "frost pockets" where nighttime temperatures often are lower than on the slopes above because cold air, like water, flows downslope. Mountainous areas have especially complex patterns of vegetation because they have such diverse and varied microclimates.

HABITATS

The effects of fire, insects and diseases, and human activity add even more complexity to vegetation patterns. A single fire can transform the plant life of an entire area. A disease can sweep through a pure stand of a single plant species like a case of measles in a kindergarten class. Chestnut blight and the Dutch elm disease, for instance, have virtually wiped out chestnut and elm trees in North America, where they once were common.

These factors produce a mosaic of distinctive habitats in every climatic region. Each plant species has its own particular environmental requirements, and it can grow only where the habitat can satisfy them. Each species has an optimum habitat in which it grows best, and it also has a range beyond which it cannot grow because of environmental limitations. Quite unrelated species of plants may have similar environmental needs and tolerances, and communities of such unrelated plants often grow in the same types of habitat.

A community of plants and its habitat form an ecosystem. The boundaries of ecosystems, like all natural boundaries, are gradual zones of transition rather than sharp lines, and they are extremely difficult to identify precisely, as anyone learns who tries to plot them on a map. Ecosystems generally are smallest and most diverse in the transition zones between major climatic and vegetation regions, where minor differences assume greater importance.

COMPETITION

Within any ecosystem the plants are fiercely competitive. The plants in pleasant wooded areas or fine flowering meadows may look perfectly innocent, but in fact they are battling fiercely with other plants for the growing space they need. Every plant must have heat and light, as well as moisture, and some plants are quite ruthless in eliminating other plants that compete with them. All plants have had to become tough competitors in order to survive. Every gardener and every farmer is all too familiar with plants so tough that they persist in growing where they are most unwelcome. We call them weeds, but one person's weed can be another person's lovely wildflower.

Plants have developed all manner of tricks for competing with and suppressing their neighbors. Some species produce dense foliage that shades out other plants. Farmers in the Midwest, for example, say their cornfields are "laid by" when the plants are lush enough to shade out most competing weeds and they no longer have to be cultivated. Some plant species blanket the ground with a thick mat of leaves that smothers the seedlings of other plants whose growing tips are not sharp enough to pierce the dense mat. The leaves and roots of some plants produce toxic substances that inhibit or prevent the growth of other species, and a few even manage to shoot themselves in the foot, so to speak, by creating conditions in which not even their own seedlings can grow.

Most plants have learned how to cope with the tricks and antics of their neighbors, but the introduction of a new species that is especially well adapted to the environment can ruin the neighborhood. In the 1930s the

Soil Conservation Service (SCS) imported from Korea a plant called kudzu and paid farmers in the South to grow it because it controlled gullies, stabilized road cuts, and could be grazed by livestock. Kudzu unfortunately went native and got out of hand. It grew everywhere. It climbed telephone poles and ran along the wires, it completely covered and smothered trees, and it threatened to blanket the entire countryside. In 1953 the SCS had to remove kudzu from the list of permissible cover plants, and in 1970 it was officially designated a weed. It still dominates extensive areas, and it seems impossible to eradicate.

Eucalyptus trees imported into California from Australia create a dense, resinous litter that inhibits the growth of other plants. Gorse and broom, tenacious weeds imported from Europe, have overrun parts of the California Coast Ranges south of San Francisco, and they must be controlled by burning, herbicides, and bulldozing to give native plants a fighting chance to grow. The problem of exotic species has become so severe that in 1988 the United States National Park Service listed encroachment by plants and animals not native to park environments as the single greatest threat to the natural resources of our national parks.

Ecology is the science that studies the physiological relationships among plants, animals, and the environments they inhabit. Each plant species has its own environmental niche, the type of habitat in which it is most comfortable and in which it can compete most successfully. The physiology of the bark, roots, leaves, and seeds of each species enables it to cope with the particular environmental stresses of specific habitats, and to compete with other species that covet the same habitat.

Thick bark, for example, protects trees against fires of low intensity. Plants that have shallow roots do better in thin soils, but they cannot tap deep sources of soil moisture, and they are more susceptible to windthrow in severe storms. Leaves that are large and fleshy promote photosynthesis, the process by which plants extract carbon dioxide from the air to make the carbohydrates they need to build tissues, and they help shade out competing plants. Small, hard, waxy leaves are better in dry areas because they reduce water loss by transpiration, and evergreen needle-leaves help plants compete on less fertile soils because they reduce nutrient turnover.

Deciduous trees and shrubs respond to winter, the cold season of physiological drought when water is frozen and thus not available to plants, by shedding their leaves in the fall and standing dormant until spring. They must produce a new set of leaves in the spring before they can start to grow, but evergreen plants can begin to grow immediately, so they get a head start in areas with short growing seasons. Annual plants complete their entire life cycles during a single growing season, and their vegetative parts disappear completely during the cold season. These plants perpetuate themselves by producing seeds that can withstand low temperatures and begin to grow when the arrival of spring ends the season of dormancy.

SEX AND SEEDS

Most plants reproduce sexually by producing seeds, although some dull species, of which potatoes, onions, sugarcane, and the grasses are examples, have little sex life. They reproduce by putting out new growth from cuttings, shoots, or runners that result in predictable clones. Sexual reproduction has the possibility of genetic mutations that can produce exciting new varieties, but it also entails the risk that animals and people can exterminate an entire species if they continue to graze, burn, or harvest the plants before their seeds have a chance to develop.

A plant primly conceals its sex organs, its stamens and pistils, in its flowers. The stamens produce microscopic spores of pollen, which insects or the wind carry to the pistils to impregnate them. The tough little bits of pollen are virtually indestructible, and palynologists study fossil pollen deposits to reconstruct prehistoric patterns of vegetation. The seeds develop at the base of the flower in the ovary, which ripens into the fruit of the plant. The seeds enable plants to reproduce, and they also enable species of plants to migrate into and colonize new areas.

Animals that eat the fruit of a plant may carry its seed considerable distances before they deposit it. Grazing cattle have been responsible for the spread of mesquite and hawthorn in two quite different parts of the United States. Shrubby mesquite bushes, which are almost impossible to eradicate, are becoming a problem on the Blackland Prairie of Texas because cattle eat the luscious pods and spread the seeds with their droppings. In similar fashion, cattle are blamed for infesting pastures in upstate New York with hawthorn bushes, which are known locally as thorn apples. The low, bushy plants have pretty white flowers and fruit that cattle like, but they form thickets so dense that people complain not even a rabbit can get through them.

Blue jays have planted lines of juniper bushes along fence rows in many parts of the South (fig. 4.2). The birds love to eat the berries of junipers, which many people call cedars. After the birds have gorged them-

Fig. 4.2.   A line of cedar trees planted along a fence line by blue jays. Cattle have browsed off the lower parts of each tree.

selves on berries they fly to the nearest fence, where they sit and preen themselves. In due time the seeds emerge, neatly encapsulated in the pellets of fertilizer they need to give them a good start in life. Birds and small mammals also help plants migrate by collecting seeds and burying them for a winter food supply. They usually forget some caches, which germinate the following spring.

Plants have other mechanisms for dispersing their seeds to new areas. Some plants, such as dandelions and maples, have seeds with hairlike or papery wings that are carried by the wind. Others, such as burs, have seeds with barbs or hooks that cling to the feathers or fur of animals. Cottonwood trees, which produce great quantities of small, light, aerodynamic seed, can spread into new areas faster than oaks and buckeyes, whose large seeds store more growth energy but depend on gravity to move them.

A British plant with a strange name, reflexed meadow grass, which is highly tolerant of salt, shows how distinctive seeds can enable a plant to colonize a new ecological niche. The natural habitat of the plant is the edge of coastal salt marshes in northeastern England, but it has spread southward along major motorways toward London (fig. 4.3). It produces large numbers of small, light seeds that lie in the mud picked up by automobiles parked near the shore. When the mud and seeds fly off the homebound cars, the seeds find a congenial home on the bare ground along the motorways where heavy winter salting against ice has killed other plant life.

SUCCESSION

Plant succession is the process by which one set of plants replaces another. It is most obvious in areas devoid of vegetation, such as bare rocks, fresh glacial deposits, open bodies of water, abandoned fields and pastures, and recently burned areas.

At one time or another every part of the earth's land surface has been bare rock, and even today plants have yet to colonize fairly extensive areas of bare rock in some parts of the world. When solid rock is exposed to the weather it begins to disintegrate into its constituent elements. The first plants that colonize a weathering mass of bare rock are small, gray, disklike lichens, which eventually create an environment in which moss can grow. The clumps of moss catch dust and mineral matter, which combine with the remains of dead moss to form a slowly thickening mat on which seed plants can begin to grow. The decomposing plant residue, called humus, which has a dark brown or black color, combines with weathered rock material to form soil. The first plants that seed the moss mat are annual weeds. They are succeeded by perennial forbs and grasses, which in turn are succeeded by shrubs and trees. This succession process may take hundreds of years, but the various stages, or seres, are clearly visible in areas where bare rock is abundant.

The plains of the Midwest north of the Ohio and Missouri Rivers were bare glacial deposits when the great Pleistocene glaciers finally melted. The first postglacial forest was dominated by hemlocks and other conifers, whose seeds were brought in by the wind. As the climate grew warmer, oaks and hickories began to grow from nuts brought in by squirrels and other small animals. Then the wind brought sugar-maple seeds and squirrels brought beechnuts, which flourished in the thick leaf mold and tolerated the heavy shade. Eventually the forest of beech and maple became dominant, because these trees cast a denser shade than oak seedlings could tolerate, and hemlock seedlings could not grow in the thick ground cover of rotting leaves.

A different kind of plant succession occurs in small bodies of water

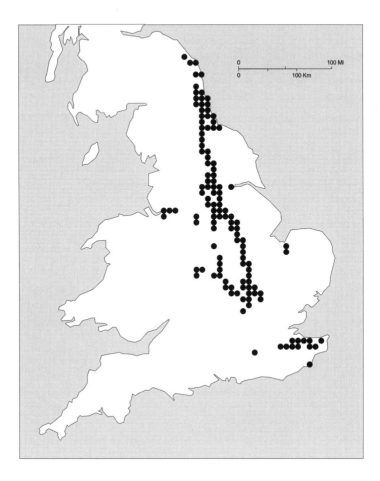

Fig. 4.3.   Distribution of reflexed meadow grass in England. Cars parked near salt marshes on the coast of northeastern England collect seeds of reflexed meadow grass, which find a congenial home where winter salting has killed other plant life along the motorways back to London. Adapted from N. E. Scott and A. W. Davison, "De-Icing Salt and the Invasion of Road Verges by Maritime Plants," *Dawsonia* 14 (1982): 42.

(fig. 4.4). Algae, water lilies, and other floating plants begin to grow in the shallow water around the edge of a lake or pond, and in time the floating mat of vegetation slowly spreads out into the open water (fig. 4.5). The plants' remains settle to the bottom, are mixed with silt washed in by rainwater, and form a sludge of peat or muck in which rushes, reeds, sedges, and cattails can grow. Eventually the soil becomes firm and dry enough to support shrubs, and finally trees. The various stages in this succession, from floating plants to marshes and peat bogs to tamarack swamps to spruce forests, may form more or less continuous rings around the lakes that dot the glaciated plains of Minnesota and Wisconsin. People who build summer cottages along the lakeshore often accelerate the destruction of the lake they love when they speed up the succession process by enriching its waters with seepage from their septic systems and fertilizer from their lawns.

Succession on abandoned fields and pastures or on burned areas is called secondary, because the invading plants are recolonizing soils that have already supported plant growth. In most of the eastern United States, annual weeds such as crabgrass take over abandoned farmland almost immediately. Perennial herbs and grasses such as goldenrod and broom sedge follow the weeds in a year or two, then shrubs start to pop up, and in thirty years or so the field will have once again become a forest of broad-leaved deciduous hardwood trees such as oak, hickory, maple, tulip poplar, and beech. The first colonizers, such as the tulip poplar, grow well in the open, but their seedlings do not like shade, and they are eventually replaced by oaks and maples, which are more shade tolerant.

The largest area of abandoned farmland in the United States is in the Southeast, where more than fourteen million acres of land, an area three-quarters the size of the state of South Carolina, went out of cotton between 1929 and 1959. The first trees to colonize former cotton land were pines, which are more tolerant than hardwoods of droughty conditions at exposed sites on bare ground. This bonanza pine forest was the resource base for the explosive growth of the pulp and paper business in the Southeast after World War II, but after foresters had harvested it, they discovered that it was naturally succeeded by hardwoods rather than by the pines they wanted. The undergrowth of the pine forest consisted mainly of hardwoods because pines need full sunlight to attain their maximum rate of photosynthesis, and pine seedlings could not compete with the hardwood seedlings, which grew better in the shade of mature pine trees.

FIRE

The foresters realized that the extensive pine forests of the Southeast had been maintained by periodic fires, some set by lightning, others by people. The Indians used to set fires to drive wild game, and the early

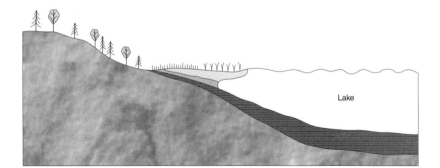

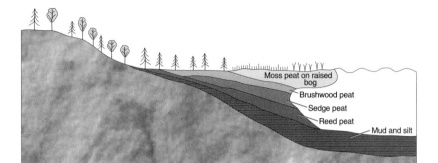

Moss peat on raised bog

Brushwood peat

Sedge peat

Reed peat

Mud and silt

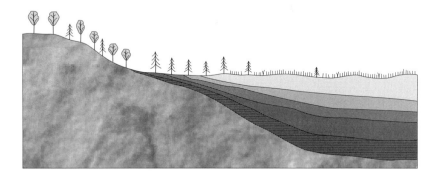

Fig. 4.4. Plant succession in a small lake. A floating mat of vegetation advances out over the water, and organic matter (peat) accumulates in the lake basin, until eventually the erstwhile lake is converted into a tamarack bog.

Fig. 4.5. A small circular lake almost completely filled with aquatic vegetation. Trees line the original shore, and the only remaining open water is the irregular dark patch. The larger lake at the top of the picture shows few signs of succession yet.

Scotch-Irish settlers were familiar with the use of fire for range management because landowners in Scotland regularly burn their heathlands to stimulate the growth of new and more palatable herbage. In the piney woods of the Southeast, people routinely set fires to kill off the undergrowth and to stimulate the growth of new grass that was more palatable and more nutritious for livestock than the coarse old unburned grasses.

"Prescribed burning" has become an essential tool of forest management in the Southeast, where approximately eight million acres of forestland are burned off each year. When weather conditions are right, foresters set low-intensity fires that they can control and keep from getting out of hand. These fires kill undesirable hardwood saplings up to three inches in diameter, but they do not harm pine trees. They prevent the buildup of a thick layer of litter that would increase the severity of a natural fire, and they expose the bare mineral soil in which pine seeds germinate best.

Natural fires have also influenced plant life in other parts of the United States, and the suppression of fires has modified the local vegetation. Periodic natural wildfires once kept trees from invading the prairie grasslands

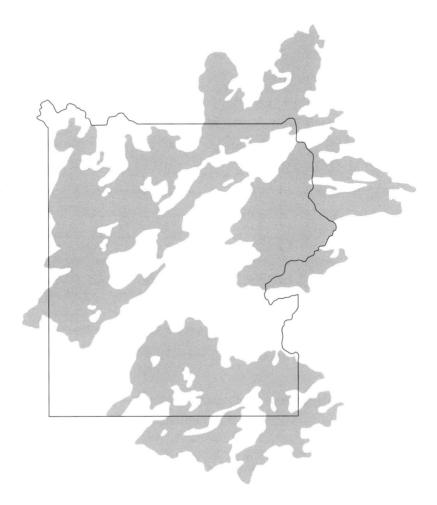

Fig. 4.6. Areas burned by wildfires in Yellowstone National Park in 1988, after Richard C. Rothermel et al., *Fire Growth Maps for the 1988 Greater Yellowstone Area Fires*, General Technical Report INT-304 (Ogden, Utah: Intermountain Research Station, U.S. Department of Agriculture, Forest Service, 1994).

of the Midwest, but fire suppression has enabled forests to spread into former grassland areas. Fire suppression in the intermontane West over the last century has encouraged the expansion of juniper woodlands, because juniper trees will encroach on and shade out adjacent grass and sagebrush areas if they are not burned every fifty years or so.

Forests of sequoia, ponderosa pine, lodgepole pine, and other fire-tolerant species are maintained in mountain areas of the West by periodic fires that wipe out competitive shade-tolerant species. For example, lodgepole pines are protected from fire damage because they have deep roots and thick bark with little resin, they grow in open stands with high, open branches, their foliage is hard to burn, and they grow best on bare mineral soils exposed to sunlight. They start to produce large quantities of viable seeds when they are only five years old, but some of the seeds are sealed to the scales of their cones by resin that does not melt until the temperature is 113°F or higher. They need the heat of a fire to open their cones and release their seeds, and fires also clear the ground of litter that would inhibit the growth of their fragile seedlings.

Sometimes lodgepole pines overdo it. They produce too many seedlings, and the densely overstocked stands grow slowly. The struggling young trees may be damaged by snow breakage, windthrow, dwarf mistletoe, and mountain-pine beetles. Dead woody fuel accumulates on the forest floor, and the lower branches of the young trees form a "fuel ladder" that enables a ground fire to climb into the canopy and produce a crown fire that destroys the entire stand. So another succession starts.

Forest managers in the West are of two minds about fire suppression. Once they stamped out all fires immediately, but people began to complain that the "unnatural" growth of underbrush was changing the appearance of the forests, and the managers started to worry that the buildup of fuel would increase the intensity of fires when they inevitably did occur. The National Park Service eventually established a policy of fighting fires started by people but letting natural fires burn. This policy was criticized severely when wildfires burned three-quarters of a million acres, an area as large as the state of Rhode Island, in Yellowstone National Park in 1988, and again in 1990, when Yosemite National Park had to be closed for the first time in history because of fires caused by lightning strikes. Still, this method seems to make more sense than any other (fig. 4.6).

CLIMAX VEGETATION

Traditional ecologists developed a theory that the final stage of succession is "climax" vegetation. Successions may begin in different habitats, or they may be initiated by different kinds of disturbances, such as fires, but within a given climatic region they gradually converge toward similar end stages, or climaxes, that are stable because they are in equilibrium with the environment. The climate is the primary environmental consideration, but distinctive topography or soils within a climatic region might result in a topographic climax or an edaphic (soil-related) climax. The climax vegetation is able to maintain and perpetuate itself. The seedlings, the young plants, and the mature plants are members of the same dominant species that is able to reproduce itself and prevent the entry of any new species. The climax vegetation is not subject to further succession unless it or its environment is disturbed, and after any disturbance it will naturally return to its previous stable state if given enough time.

Like any good theory, the theory of a climax vegetation concentrates on the interaction of a restricted number of variables to help us make sense of the complexity of the real world we live in, and like all theories, it has its limits. Certainly climatic patterns clarify the broad general patterns of vegetation at the global and regional level, but at the detailed local level the vegetation is a mosaic of many variables that are rarely stable. The vegetation of any area is regularly subject to such disturbances as long-range climatic change, annual and seasonal fluctuations in weather, insects and diseases, high winds and fires, and especially the profound effects of human activities.

THE MANAGEMENT OF VEGETATION

Our world is constantly changing, and there is no "natural" state of equilibrium to which it will return if any particular disturbance is controlled. There probably is no plant community anywhere in the world that has not been affected in some way by human activity, so there is no such thing as truly natural vegetation that has not been modified in some way by people. In North America we sometimes talk about the "contact" vegetation, which is the vegetation that was first seen and recorded by white explorers, but this contact vegetation had already been greatly modified by centuries of occupance by Indians before the first white people ever arrived.

Some environmentalists have argued that the vegetation of certain areas should be returned to what they suppose is its natural state, but the theory of climax vegetation is not a sound basis for any resource-management policy. The goal of resource management should be the identification of the human interventions that should be promoted or opposed to achieve some desired condition, and not an attempt to return to some theoretical natural equilibrium.

An understanding of the processes that influence the growth of plants is essential to any attempt to manage vegetation. Nature often fails to provide what people want or need from it, and people must manage nature to satisfy their requirements. The secret of managing nature is to develop the fullest possible understanding of natural processes, and then to manipulate or modify these processes to achieve the results desired. Tampering with nature is always risky, however, because natural processes are so complex that people rarely understand them completely, and many attempts at natural resource management have had unintended, unanticipated, and sometimes unappreciated side effects.

 MOUNTAINS, CLIMATE, AND VEGETATION

Mountainous areas have complex patterns of vegetation because they are cooler and wetter than the adjacent lowlands. They are cooler because the atmosphere is heated by the surface of the earth, not by direct sunlight. Direct sunlight heats the surface, which then radiates heat to the air above it. The air closest to the surface usually is the warmest, and the temperature of the air decreases with height above the surface. The average decrease per thousand feet is 3.3°F, but it is 5.5°F in dry air and only 2.2°F in moist air, because moisture conserves temperature.

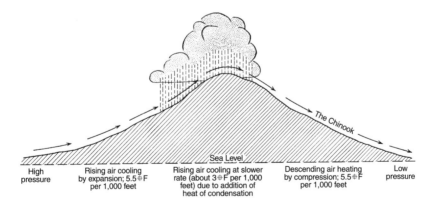

| High pressure | Rising air cooling by expansion; 5.5⊕F per 1,000 feet | Rising air cooling at slower rate (about 3⊕F per 1,000 feet) due to addition of heat of condensation | Descending air heating by compression; 5.5⊕F per 1,000 feet | Low pressure |

Fig. 4.7. The influence of a mountain range on precipitation, with heavy precipitation on the upper slopes on the windward side and a rain shadow on the leeward side. Reproduced by permission from Vernor C. Finch and Glenn T. Trewartha, *Physical Elements of Geography*, 3d ed. (New York: McGraw-Hill, 1949), 214.

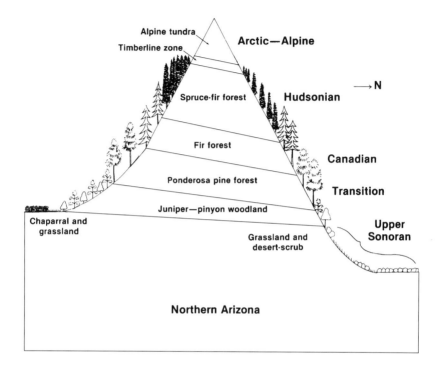

Fig. 4.8.   Vertical zonation of vegetation on a mountain range in northern Arizona. The belts of vegetation are at lower elevations on the cooler north-facing slope. Reproduced from David E. Brown and Charles H. Lowe, *Biotic Communities of the Southwest*, General Technical Report RM-78 (Fort Collins, Colo.: Rocky Mountain Forest and Range Experiment Station, U.S. Department of Agriculture, Forest Service, 1980).

In the middle latitudes air masses usually move from west to east. A moist air mass moving eastward from the Pacific Ocean has to rise when it crosses a mountain system, and as it rises it cools at the moist-air rate of 2.2°F per thousand feet. Eventually it cools to the dew point, the temperature at which its moisture condenses and falls to the ground as rain or snow, and that is why the windward flanks of mountains usually have heavier precipitation and lusher vegetation than do the lowland areas around them (fig. 4.7).

The downwind, or leeward, sides of mountain systems are quite another story. The mountains milk the moisture from the air, and the dried air warms up at the 5.5°F rate when it descends on their downwind sides. The dry descending air acts like a sponge that absorbs moisture from the land instead of producing any precipitation. A mountain system thus acts as a kind of umbrella that keeps precipitation away from the areas downwind, to the lee of the mountains, which lack moisture and are said to suffer from a rain-shadow effect.

Lower temperatures offset the advantages of greater precipitation at higher elevations in the mountains, and broad horizontal bands of vegetation reflect these variations in temperature and precipitation (fig. 4.8). The lower slopes are wooded, but the trees begin to thin out toward the timberline, and the higher areas above the timberline have alpine meadows much like the arctic tundra. These bands are higher on the south-facing slopes, which are warmer and drier because they are tilted toward the sun, and lower on the cooler, moister, shadier slopes that face north.

Mountainous areas also experience extreme forms of air drainage. Cold air is denser and heavier than warm air, and on calm, clear nights it flows invisibly downslope to puddle up in the valley bottoms and other lowland areas, which have the lowest air temperatures and the greatest risk of frost. Even in areas of low relief, farmers have learned to avoid these lowland frost pockets, and they plant their orchards, vineyards, and other delicate crops on sloping land to reduce the danger of frost damage.

5 *The Use of Forests*

The growth of plants is so intimately related to the physical character of the sites on which they grow that the existing vegetation neatly summarizes all the ecologically relevant variables, and it is the most readily visible indicator of environmental conditions. Cactuses do not grow in marshes and cattails do not grow in deserts. Other plants may send out more-subtle signals about their habitats, but all plants can "speak" to the person who has learned to read their signals aright. They tell a tale of sunshine and shadow, of wind and rain, of soils that are poorly drained or droughty, that are richly or poorly endowed with the essential plant nutrients. They also tell the tale of how people have used and abused them.

The early white settlers of North America were full well aware that the existing plant life was an excellent indicator of the environmental potential of new and unfamiliar areas, and they used vegetation as their guide to tell them which land to choose and which to avoid. Handbooks for immigrants that were published in Europe in the nineteenth century discussed at great length the value of tree species as indicators of soil quality in North America. In general, they said, deciduous forests indicated better soil than coniferous forests, and forests with many species signaled better soil than forests with just a few. Maple, elm, ash, walnut, and hickory indicated the best soils, they said, and they warned that pine land was poor land.

The Pennsylvania German farmers who migrated to Canada after the Revolutionary War used black walnut trees as their guide in selecting land, because they knew that these trees grew on the fertile limestone soil they sought. The gentle swells and swales of the glacial plains in Rush County, Indiana, are almost imperceptible to the human eye, but the vegetation at the time of settlement clearly showed their differences. The first settlers avoided the "beech land" in the swales, where water stands late into the spring, but they eagerly claimed the "sugar-tree land" on the swells, which was dominated by sugar maple and oak, because they knew it was well drained.

Some plants signal quite specific environmental conditions. In east Texas, for example, blackjack oak indicates dry, sandy soils with low organic content, and overcup oak indicates poorly drained clay soils. In southern Illinois, pin oaks grow in nearly pure stands called "pin oak flats" on heavy claypan soils that are poorly drained and damp in winter and spring, but so dry for part of the summer that other trees cannot compete. In most of the eastern United States the Eastern red cedar seems to be nature's own designated environmental indicator for droughty soils, especially thin soils underlain by limestone (fig. 5.1). The "cedar glades" of the Nashville Basin in central Tennessee are almost-pure stands of red cedar on fairly level areas of bare or nearly bare limestone.

THE TREELESS PRAIRIES

Perhaps the best known and most vigorously debated vegetation indicator in the United States is the prairie grasslands that once formed a wedge widening westward from the Indiana-Illinois line across the rolling glacial plains of the Midwest. The prairies have enough precipitation to support tree growth, but when white explorers first saw them they had no trees, only knee- to waist-high grasses. The grasses added humus to the soil each year. The combination of adequate precipitation and deep, dark, humus-rich soil made the prairies one of the finest farming areas in the world, but the early settlers called them "the barrens," because the English language does not have a good word for a vast extent of treeless grassland. The French word *prairie* eventually replaced "barren," but old-timers in Iowa still refer to some of the world's finest farmland as the barren prairie.

Some people suppose that the early settlers in the Midwest avoided the treeless prairies because they needed wood to build their houses, their barns, their fences, and most important of all, their fires for cooking and heating. It is clear that the first settlers chose sites in the parklike "oak openings" that bordered the prairies, and those who settled on the open prairie acquired distant woodlots in the wooded areas that fringed the prairie streams, but the real reason they bypassed the prairies was their inaccessibility, not their treelessness. Travel by water was easier than travel overland in the days before roads were built, and the prairie areas simply were too far from the navigable rivers.

It is difficult to test the idea that less-productive land near a navigable stream was settled before better land that was less accessible, because the distribution of woodland and prairie accords so closely with the flow of

Fig. 5.1.   Eastern red cedar saplings on an area of thin soils underlain by limestone.

migration. The pioneers moved from east to west, from woodland to prairie, and their principal routes were the rivers whose valleys sheltered the westernmost tentacles of woodland. The early settlers certainly used the wooded valleys to bypass upland prairies, but nowhere did they leapfrog directly across a prairie to reach a wooded area on the farther side, and in at least one area, in southeastern Minnesota, upland prairies were settled before wooded valleys because a military road had been built across the prairies. The extensive prairie areas were not really settled until the construction of railroads made them accessible. Small towns developed along the railroad lines, and in each small town the lumberyard and coal dealer beside the tracks provided the building material and fuel that settlers farther east had gotten from their own woodlots.

THE EARLY USE OF FORESTS

The prairie areas of the Midwest have excited attention in part because, unlike most of the world's great grassland areas, they are fairly densely populated. People generally avoid grasslands because they are too dry. Most of the world's people live in areas that once were forested, because trees are the dominant form of plant life in areas that enjoy enough precipitation to grow the principal cultivated food crops.

The original forests have long since been cleared so the land could be used for crops. The first settlers cut down most of the trees; they set fire to the woods that remained, to drive out wild animals and to improve the forage for their cattle, sheep, and goats. These domestic animals have effectively prevented the regrowth of trees by nibbling, trampling, and destroying the seedlings and young shoots. Deforestation has had devastating side effects in some parts of the world, especially in hilly areas where trees protect the steep slopes by intercepting precipitation, slowing down runoff, and reducing the danger of soil erosion. The removal of trees has exposed the topsoil to accelerated erosion, and the eroded topsoil that was carried down to adjacent lowlands has clogged streams and transformed the lowlands into malaria-ridden marshes.

Americans have been singularly cavalier about their forests, which once seemed both inexhaustible and interminable. A Scotch-Irish frontiersman, it was said, could not see a tree without wanting to cut it down, and many people once treated trees like weeds that had to be removed before the land could be used to grow crops. In fact, the forests provided many things the early settlers needed. First and foremost was firewood for heating and cooking. Wood still provides 90 percent or more of the fuel that is used in many Third World countries, and some of them are facing critical shortages. In Ethiopia, for example, 40 percent of the land was forested in 1900, but by 1990, the figure was a mere 3 percent, and people had to travel enormous distances in search of firewood.

The forest provided the logs and lumber that the early settlers used to build their houses, and the shingles for their roofs. The pioneers used wood for virtually everything that is made out of plastic today: furniture, dishware, containers, handles for tools and implements. They cut posts and split rails to fence their crops against marauding animals. They learned which nuts, fruits, and berries were safe to eat, they followed bees to the hollow trees where their honey was cached, and in the spring they tapped the rising sap from maple trees and boiled it down for syrup. They used the bark of trees for dyes, and from it they extracted the tannin they used to cure animal hides into leather. They saved wood ashes in hoppers, dripped water through them to leach out the potash, and mixed it with animal fat to make soap.

When Americans began to put out to sea they used wood to build their ships for fishing, for trading, for whaling, and when necessary, for fighting. At one time nearly every coastal town in New England had its own shipyard. The early sailing ships needed wooden planks for their hulls and decks, and each one needed more than twenty masts and spars that were long, straight, strong, durable, and flexible. The largest ships had mainmasts forty inches in diameter and 120 feet long. The majestic white-pine trees of Maine, still the Pine Tree State, made such superb mainmasts that in 1722 the English government passed a law reserving all white-pine trees for the Royal Navy, but the New England colonists thumbed their noses at the royal decree.

Until the Civil War, steamboats and mining consumed large quantities of wood. People who lived near navigable rivers and lakes could stack wood on the banks and sell it to the passing steamboats, which had to stop twice a day to take on fuel, of which they consumed prodigious quantities. Mining areas also needed large amounts of wood, strong pit props to shore up the roofs of the mines, and charcoal to smelt the iron ore and other minerals they produced. Many areas had small charcoal blast furnaces that used deposits of low-grade iron ore to make iron for tools, implements, rifles, and other essential metal goods.

NAVAL STORES

Among the first commercial products of the forests of the United States were naval stores (turpentine, resin, tar, and pitch), so called because back in the days of wooden sailing ships they were used to caulk the hulls and to protect the ropes and rigging. As Michael Williams noted in *Americans and Their Forests*, as early as 1608 a boat loaded with "pitch, tarre, clapboard, and waynscot" sailed for England from Jamestown, Virginia. The first major area to produce naval stores was the sandy Coastal Plain of eastern North Carolina, which still rejoices in its nickname, the Tarheel State, but since 1900 the principal area of naval-stores production has shifted southward to the piney woods of northern Florida and southeastern Georgia, which account for about two-thirds of the total world supply.

In March, when the sap starts to rise, workers shave off a diagonal strip of bark about a third of the way around the base of each tree (fig. 5.2). They nail a metal collecting gutter across the bottom of the shaved area, and nail a two-quart collecting cup at the lower end of the gutter. Then they chip off an inchwide streak of bark above the gutter, and the sap slowly oozes down

Fig. 5.2.   Southern pine forest being tapped for naval stores.

into the cup. They chip new streaks every two weeks from March to November, and every month or so they collect the gum from the cups and haul it to a turpentine still to be refined and prepared for shipment to market.

The life of a shaved face is five or six years, and then the workers chip the other side of the tree, leaving four-inch bark bars between the two faces as "life lines" through which the tree can receive moisture and nutrients from the soil. A pine tree stops growing when it is chipped for naval stores, and it is cut down and sold as soon as its faces are finished. Sawmill operators do not like logs that have been chipped for naval stores, because some of them still contain nails that can tear out the teeth of a saw.

COMMERCIAL LUMBERING

Small-scale lumbering was closely associated with the spread of settlement in the eastern United States. Each newly settled area needed a combined gristmill, where wheat could be ground into flour, and sawmill, where the trees cleared from the land could be turned into boards, planks, tool handles, barrel staves, and all the other wood products the settlers

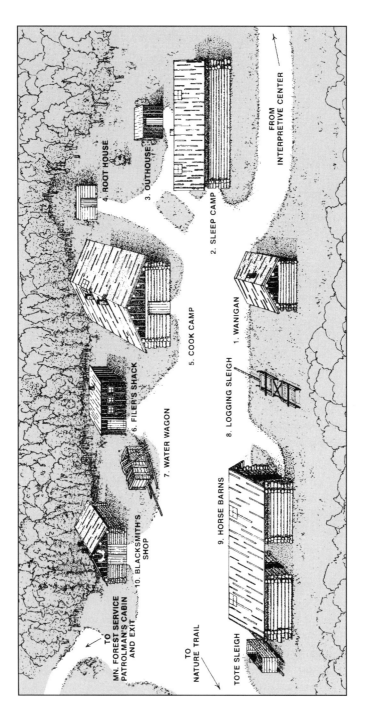

Fig. 5.3. The site map of a reconstructed late-nineteenth-century logging camp at the Forest History Center near Grand Rapids, Minnesota. Reproduced by permission of the Minnesota Historical Society.

needed. Most of these mills were small, at power sites on streams, and they served local needs, with little commercial surplus.

Commercial lumbering on a large scale first developed in the upland areas of New England, New York, and Pennsylvania, where much of the land was not well suited to farming. The sawmills were at power sites on rivers. Initially they used logs that were hauled in from the local area, but when it had been logged off the lumbermen began to use the river to float in logs from much greater distances. They cut down trees in winter, used teams of oxen or horses to skid the logs down to the nearest stream, and stacked them until the spring thaw. Each owner had a unique log mark, like a cattle brand, that the lumberjacks hammered into the end of each log so it could be identified at the boom area.

The annual spring log drive to the sawmill was an exciting and dangerous affair, because the lumberjacks could be crushed in logjams or drowned. The logs careened downriver to the quiet waters of the boom area upstream from the sawmill, where the river was blocked and divided into pockets by floating booms of logs chained together. The logs were released slowly through a gate in the main boom, and the "catchmarkers," who knew every log mark on the river, directed them into the pocket that belonged to their owner.

THE BOREAL FOREST OF THE GREAT LAKES STATES

By 1840 the lumbermen had so depleted the forests of the Northeast that they began to carry their skills westward to exploit the boreal forests of northern Michigan, Wisconsin, and Minnesota. The first operations were small, and the loggers still used river drives to take logs to the sawmill. In the fall they built splash dams on small streams, and in winter they stacked logs on the ice that formed on the temporary lakes. In spring they dynamited the dams, and the logs went cascading through. Later they built permanent dams with sluice gates that could be opened to release the logs.

After the Civil War, the lumber companies began to use railroads to transport logs from areas inaccessible by water, and within half a century they had logged off virtually the entire boreal forest around the upper Great Lakes. They used power-driven machinery, and they enlarged their operations. The early logging camps had simple log cabins for ten to twenty-five workers, but the newer and larger camps housed fifty workers or more in rough lumber and tarpaper buildings (fig. 5.3).

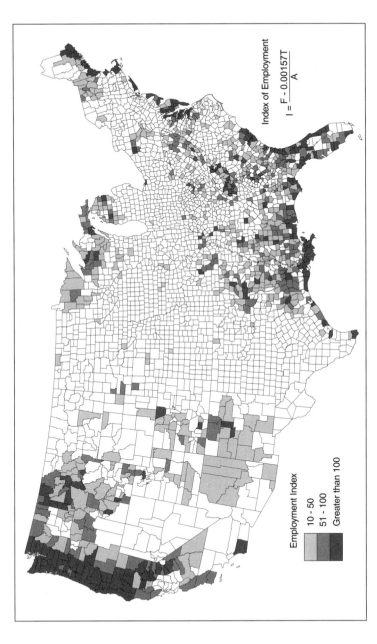

Index of Employment

$$I = \frac{F - 0.00157T}{A}$$

Employment Index

10 - 50

51 - 100

Greater than 100

Fig. 5.4. An index (I) of employment in forestry and fisheries in the counties of the United States in 1980. F is the number of persons employed in forestry and fisheries, T is the total number of persons employed in the county, and A is the area of the county in hundreds of square miles. The Census of Population published by the U.S. Bureau of the Census does not report separate data on the numbers of persons employed in forestry and in fisheries, so the two must be combined.

Many of the lumberjacks were husky young men from farms on the prairie who wanted to make extra money by working in the woods in winter. On Saturday night, when they sought excitement in the nearest town, the town fathers often rolled up the sidewalks, which were made of boards wired together, to protect them against being mangled by the steel-calked boots of the roistering lumberjacks. The towns were one-industry communities that depended entirely on their mills. They had a few company houses for managers and married workers, but their boarding houses, company dormitories, bars, and bordellos indicated that most of the workers were young, single, transient males.

Few logging camps were used for more than three years, and the minor logging railroad lines were abandoned and dismantled when an area had been logged out. The mills were only slightly more permanent. They were closed and torn down when logging ended, and their equipment was moved to newer and larger mills closer to the source of timber. Today the boreal forest around the upper Great Lakes is scored with abandoned railroad lines (some of which have been converted into recreational trails), it is littered with what remains of old logging dams, and it is dotted with ghost towns, sites once occupied by logging camps or mill towns of which little remains but earth-covered mounds and traces of a few foundations.

The lumber companies, the railroad companies, and even some state agencies tried to entice farmers to settle in the cutover areas after they had been logged off, but the climate was too harsh and the soil too infertile for successful agriculture. Unintentionally, however, lumbering laid the foundation for the development of the tourist industry. The construction of a network of railroads and later truck roads for logging opened up large areas for people who come to hunt, fish, camp, canoe, and seek other forms of recreation in remote areas that would otherwise be inaccessible.

THE PACIFIC NORTHWEST

The boreal forest of the upper Great Lakes had been completely cut over by 1910, and once again the lumbermen moved on, some to the Pacific Northwest, others to the South (fig. 5.4). The majestic forests of British Columbia, Washington, and Oregon have become the leading lumber-producing area of contemporary North America. The lumberjacks live in towns and commute to their jobs in the woods in their pickup trucks. Much of the forestland is publicly owned and managed by the United

States Forest Service, which restricts the size of individual clearcut areas, so the land is a crazy quilt of logged-off patches separated by narrow strips of trees planted to reseed the cleared areas and to serve as firebreaks between them.

The lumberjacks cut the felled trees into thirty-three-foot logs to be hauled to the mill by logging truck or railroad. In remote areas along the coast they haul the logs to protected bays and assemble them into enormous rafts that powerful tugboats can tow through sheltered coastal waters to tidewater mills. Modern forest industry complexes in the Pacific Northwest include plywood mills, pulp and paper mills, and other woodworking mills, in addition to large sawmills, to enable them to use completely and efficiently the great variety of trees that come in from clearcut areas.

MODERN FORESTRY IN THE GREAT LAKES AREA

In the upper Great Lakes area some forest industry companies managed to buck the trend to "cut out and get out" after World War I by learning to use the second-growth forest of smaller aspen and birch trees that grew on the cutover areas. Aspen and birch, which were considered weed trees in the lumber boom days, have become the basic raw materials of a new forest-products industry that produces pulp and paper, hardboard, waferboard, particleboard, packaging, posts, poles, and even chopsticks.

Logging has become completely mechanized (fig. 5.5). As Clayton Rollins, a logging contractor near Pine River, Minnesota, recounted:

> My father had thirty men working for him, and he expected each one to cut a cord of wood a day. Three men work for me, and we cut forty-fifty cords a day easy. We use machinery with a replacement cost of half a million dollars, but I'd say you're not overloaded if you've got a million dollars worth of logging equipment. We start work at 6, but I've got to be out there by 5:30 to start the machinery and get it warmed up for the day's work. We just stay home when it gets down below twenty-five below.
>
> We stop at 3:30 to get things fixed up for the next day's work, and to be sure there are no fires before we leave. We take off the first week of deer-hunting season to keep from getting shot at, and we shut down for five or six weeks during the spring thaw because weight restrictions on the highways won't let us use fully loaded trucks, and you're just wasting money if you send a half-loaded truck to the mill.

Fig. 5.5. A sketch of a modern whole-tree wood-chipping operation. A feller-buncher seizes a tree and snips it off at ground level with powerful hydraulic shears. A skidder tractor drags bundles of trees to the chipping unit, which blows the chips into the van that hauls them to the paper mill. Reproduced by permission of the author, James A. Mattson, North Central Forest Experiment Station, U.S. Department of Agriculture, Forest Service, Houghton, Michigan.

He can fell 100 to 140 trees in an hour with a feller-buncher, a small but powerful tractor with a pair of hydraulic shears mounted at ground level on the front. It has steel claws that seize and hold the trunk of a tree while the shears cut it off; they can snip through a twelve-inch pine tree as easily as scissors cut through paper. The claws then twirl the severed tree like a baton and place it in a pile with a couple of other trees for a tractor to skid to the harvesting machine. The harvester strips off the limbs, saws the trunk into eight-foot logs, and stacks the logs beside the road to be picked up by the truck that will haul them to the mill.

THE SOUTHERN PINE FOREST

By 1890 some lumbermen had begun to realize that the days of the Great Lakes forest were numbered, and they moved their operations to the pine forests of the South. By 1930 they had pretty well denuded the region,

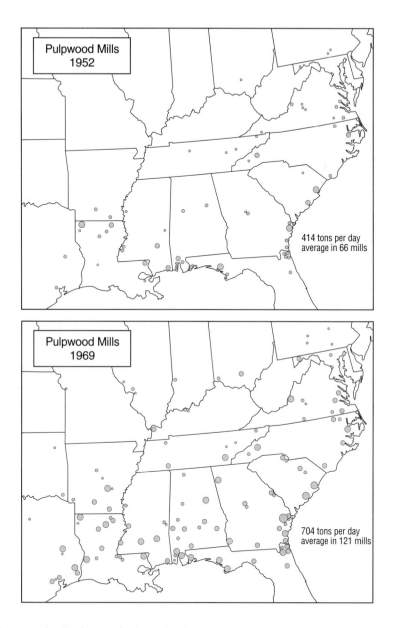

Fig. 5.6.   The distribution of pulpwood mills in the South in 1952 and in 1969. The pattern and capacity of mills have changed only slightly since 1969.

Fig. 5.7. Piles of logs waiting to be processed at the forest-products mill in the background.

and their sawmills and company towns disappeared almost as quickly as they had appeared. For two decades, small, portable, "peckerwood" sawmills picked over the second growth from the cutover areas, but since World War II forest industry companies have turned the South into the nation's leading pulp- and paper-producing area, first using the bonanza pine forest that flourished on abandoned cotton land, then plantations of planted pine trees that grow rapidly in the region's subtropical climate (fig. 5.6).

Wood consists of cellulose fibers that are held together by a gluelike substance called lignin. At a pulp mill the wood is chipped into small pieces and boiled with strong chemicals to remove the lignin and leave a soft, soupy pulp of cellulose fibers. The pulp is poured onto a continuously moving wire screen that drains off the water, and then the wet pulp is squeezed through a set of rollers and driers, from which it emerges as an endless sheet of paper.

Disposing of the strong acid waste containing dissolved lignin from pulp and paper plants is extraordinarily difficult. The dumping of hot acid waste into streams will kill fish and plant life. The release of chemical fumes from tall stacks gives pulp and paper mills an unforgettable acrid smell that makes your eyes water, burns your nose, and flakes the paint

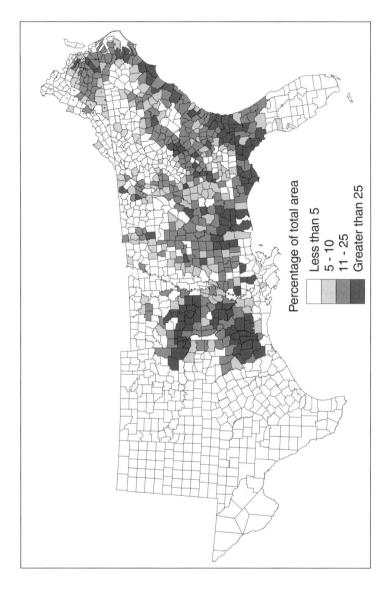

Fig. 5.8. The percentage of the land in each county in the South that was owned by forest-industry companies around 1985.

off houses. The exits from the parking lots at many mills have automatic sprinkler systems that rinse off the corrosive waste that has settled on cars parked in the lots.

Enlightened pulp and paper companies have made herculean efforts to control and reduce pollution, but some companies are reluctant to make massive investments in nonproductive plant and equipment changes, and their workers are understandably reluctant to complain, because they fear the loss of jobs in areas where jobs are scarce. As one of them said to me, "You may not like it, but that smell is the smell of jobs and money, and there ain't much else of either one around these parts."

The forest-products industry has been moving steadily toward larger, more-efficient, fully integrated mill complexes, and it is constantly developing new technologies that will enable it to use as much as possible of everything that comes out of the woods (fig. 5.7). The waste from one mill becomes the raw material for another, or it is burned to raise the steam that generates electricity for the entire complex.

For example, Crossett, Arkansas, a town of sixty-three hundred people, has a forest-products-industry complex that employs more than three thousand workers in a lumbermill, two paper mills, two plywood mills, a particleboard plant, a tissue mill, a bag plant, an extrusion plant, and chemical plants that make formaldehyde, resin, wax emulsion, tail oil, dried soap, and turpentine. The company has impounded a seventeen-hundred-acre lake to ensure a dependable supply of water for the complex. It has stocked the lake with fish and developed a public lakeshore park with charcoal grills, a picnic shelter and tables, restrooms, and trailer hookups.

Forest-industry companies in the South have acquired vast tracts of forestland to ensure a steady supply of raw material for their mills (fig. 5.8). In the area from eastern Oklahoma and Texas to Virginia they have bought forty million acres and taken long-term leases on another four million, a total area larger than the entire state of Georgia. Each year they plant young pine trees on more than half a million acres, much of it land from which the natural pine has been cut, and they manage it intensively both for maximum production and as an example for the owners of neighboring woodland areas.

Most of the woodland in the South, the nation's most heavily wooded area, still belongs to private citizens who have diverse reasons for owning it. Many are not particularly interested in cutting their trees; their tracts are too small and too scattered for good forest management, and even those who would like to sell pulpwood may not be able to afford the expense of

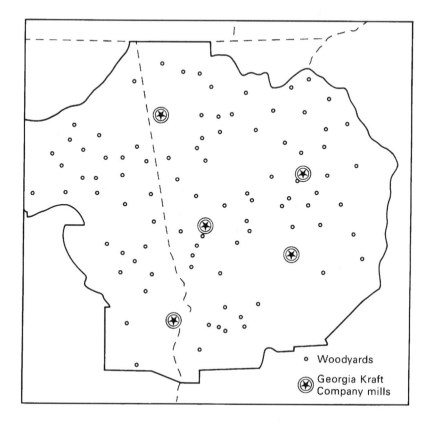

Fig. 5.9. Woodyards and processing mills in the operating area of the Georgia Kraft Company in northern Georgia and northeastern Alabama. Reproduced by permission from *Annals of the Association of American Geographers* 70 (1980): 514.

transporting it to a distant mill complex. The forest-products industry has tried to encourage pulpwood production by creating collection points known as woodyards where local landowners can sell their pulpwood; there it is combined into large loads for shipment to the mills. Today woodyards are scattered across the South at intervals of fifteen to twenty-five miles (fig. 5.9).

6 *Cropping Systems*

Crop selection is not a stochastic process. Farmers do not simply flip a coin to decide which crops to plant next year. Their decisions are constrained by economic, ecological, and a host of other considerations. Most farmers probably want to plant the crop that will make them the most money, based on their knowledge of the costs of producing specific crops and their best guesses about what the price is likely to be when they are ready to sell.

Farmers realize, however, that they may be asking for trouble if they plant the same crop year after year on the same ground. Concentration on a single crop, or monoculture, enables farmers to hone their management skills, reduce their costs of production, and facilitate marketing, but bad weather or low prices can be disastrous for those who have put all their eggs in one basket by growing only a single crop, and a new insect or disease can sweep through their fields like a case of measles in a kindergarten class.

Most farmers have additional reasons for wanting to grow a variety of crops. They like to spread their work through the year as evenly as possible instead of having it all pile up at short but ferociously busy peak periods for planting and harvesting a single crop. Diversifying the cropping system also can reduce the risk of erosion, and it can use the fertility of the soil efficiently because different crops remove different nutrients from the soil. Alternating deep- and shallow-rooted crops can equalize the removal of nutrients from different soil horizons.

Some crops actually add nitrogen to the soil, and farmers can plow under the leaves, stalks, and stems of plants to add organic matter to it. Organic matter, or humus, keeps the soil loose and friable and maintains its physical structure and porosity. Sandy soils that lack humus are droughty and easily leached. Clay and loam soils that lack humus are sticky, harder to plow, and more easily compacted.

Weeds are fairly easy to control in some crops, but virtually impossible in others. Some crops must be protected by a hardier "nurse crop" when they are young and delicate. A given crop may not yield well after another crop, but yield quite well after a third. For example, farmers who grow sugar beets, potatoes, oats, and alfalfa in irrigated oases in the West have learned that sugar beets yield much better after potatoes than after either oats or alfalfa, but potatoes do not seem to care.

CROP ROTATIONS

In many parts of the world farmers have developed cropping systems, or crop rotations, that have persisted for decades or even for centuries because they are economically successful and ecologically sound. A crop rotation is a cropping system in which farmers continue to grow the same crops in the same sequence on the same piece of ground. Crop rotations are flexible, and farmers will modify them on individual farms or even on individual fields, but these modifications are merely variations on the customary crop rotation of the area. Crop rotations are the key to understanding the patterns of crops grown in various parts of the world.

Farmers group crops into three broad categories when they plan rotations: cultivated row crops (including fallow land), hay or sod crops, and close-sown small-grain crops.

Row crops, such as corn, cotton, fodder roots, potatoes, sugar beets, soybeans, and vegetables, are planted in rows that can be cultivated. The verb *to cultivate* has both a general and a quite specific meaning. Generally it simply means "to grow," as in the sentence "Farmers in the Midwest cultivate corn and soybeans," but when farmers talk about cultivation they specifically mean "to loosen or break up the soil in order to kill weeds before their seeds have time to mature, and to increase the water retention of the soil." Farmers can chop out weeds with hand hoes, or they can uproot and bury the weeds with special implements pulled by horses or tractors. A modern, tractor-drawn, twelve-row cultivator with multiple spring tines that claw the soil between rows of plants is a far cry from a hand-held hoe.

Most row crops are cash crops, and a row crop usually is the first, most important, and most profitable crop, around which the rest of the rotation is organized. Normally it gets most of the fertilizer, and the less valuable crops that follow it must depend on the residual fertility in the soil. The big drawback of row crops is that they encourage erosion, especially on sloping land, because the bare ground between the rows provides a ready channel for rainwater to run off rapidly, and it carries away the finer soil particles with it.

Hay or sod crops, such as alfalfa, clover, and a wide variety of grasses, can help reduce soil erosion because they slow down the runoff of rainwater. As the saying goes, "Hay crops in a rotation are like brakes on an automobile; the steeper the slope, the more you need." They usually provide the least direct income because they are grazed by livestock or cut for hay, but they can enrich the soil by adding humus to it. Leguminous hay crops are especially helpful because they can fix nitrogen by extracting it from the air and storing it in their roots, which remain in the soil to fertilize the crop that follows. They will fix only as much nitrogen as they need, however, so it is not a good idea to grow them after row crops that have been fertilized heavily. Good hay crops are hard to establish, and during their early growth stages they benefit from the protection of a small grain as a nurse crop.

The close-sown small-grain crops (wheat, rye, barley, and oats) are good nurse crops for young hay plants, but they are almost equally effective at sheltering weeds, and they must be rotated with other crops to get rid of the weeds. The small grains were the earliest domesticated plants, and for most of human history they were the only field crops, as distinct from crops that were grown in small garden plots. The small grains were the basis for the "three-field" agricultural system that persisted from Roman times through the Middle Ages in Europe north of the Alps.

FOOD, FEED, AND FALLOW

The medieval three-field (food-feed-fallow) rotation was based on a food grain (wheat or rye) in the first year, a feed grain (barley or oats) in the second, and bare fallow in the third, when the farmers plowed the field periodically to kill the weeds. One-third of the land had to be left fallow simply to control weeds, and that land produced no crops. Wheat and rye need long growing seasons to produce good yields, and the farmers planted them in the fall on the field that had lain fallow all summer. They were harvested at the end of the following summer, and during the fall and winter, when weather conditions permitted, the farmers plowed the fields in which they had grown to prepare the soil for spring planting of the feed grain, barley or oats, which ripened for August harvest. After harvest the barley or oats stubble was grazed by livestock over winter and plowed up in early summer; then the field lay fallow until it was planted with the bread grain in the fall, and a new rotational cycle began.

Wheat, the preferred bread grain, and barley, the better feed for livestock, were grown on the productive loam soils of the loess lands of western Europe, but the peasants on the poorer, sandy soils of the great glacial outwash plains of eastern Europe had to make do with rye bread on their tables and oats for their livestock. The basic medieval rotation was wheat-barley-fallow on loam soils and rye-oats-fallow on sandy soils.

ROOTS AND CLOVER

The introduction of new row crops, such as the lowly turnip, and new leguminous hay crops, such as clover, changed the medieval three-field system into a new four-field system and launched the Agricultural Revolution. The fallow year was no longer necessary; farmers could cultivate the row crop to eliminate weeds, and the leguminous hay crop enriched the soil. Bringing the third, or fallow, field into production each year increased the acreage of land under crops by 50 percent, and the new crops, which produced more and better winter feed for livestock, boosted agricultural production still further. The Agricultural Revolution had to precede the Industrial Revolution, because it enabled farmers to feed the rapidly growing urban population.

A better supply of winter feed enabled farmers to keep more livestock, and it set in motion an upward spiral: more feed, more livestock; more livestock, more manure; more manure, richer soils; richer soils, still more feed. Turnips were important in the early years, but they are a fodder root of relatively low value, and many farmers replaced them with cash crops such as sugar beets or potatoes. The standard four-year rotation in contemporary Europe is wheat–sugar beets–barley–clover on the better (loam) soils and rye-potatoes-oats-clover on sandy soils.

CORN, BEANS, AND CUCURBITS

The first European settlers in North America encountered a strange new crop: corn. They adopted the Indian method of growing it with beans and cucurbits (pumpkins, squash, gourds, cucumbers, melons) in small patches of cleared land that produced all the food they needed. The tall stalks of corn supported a variety of climbing bean plants, and the cucurbit vines crept across the ground beneath them.

Corn was the staple food. The pioneers boiled fresh corn and beans to make a tasty dish of succotash. They leached dry corn in lye to make hominy; they ground it into cornmeal that they baked on the hearth or fried in a skillet. The pioneers also grew a bit of tobacco, perhaps some potatoes, grain sorghum to make molasses for sweetening, and cotton, flax, and hemp for fibers they could make into homespun clothing. In New England, farmers used the marshes as cattle pastures and cut marsh hay for fodder, but the early settlers in Appalachia kept few livestock because they got most of their meat by hunting wild game. After they had decimated the wild-game supply they turned cattle and hogs loose to run semiwild in the woods, and often they hunted them just as they had hunted wild game.

Their small cleared patches of cultivated land were scattered through the forest, and farmers in the South continued this traditional patch agriculture even after they started to grow commercial crops such as cotton and tobacco. They rotated land rather than crops because maintaining soil fertility is so difficult in a hot, humid, subtropical area. They cultivated a patch until they had exhausted its fertility, and then they cleared a new patch and abandoned the old one to grow up in brush and eventually trees. A patch might be cleared once again after it had rested under trees for fifty years or more and regained some of its fertility, but few people lived long enough to see the same piece of ground cleared twice in a lifetime.

This system of shifting cultivation, which Southerners prefer to call "brush fallow" or "land rotation," created a landscape that deceived even knowledgeable observers. One-time visitors reported that much of the land was abandoned, as indeed it was, but only for a generation or so. Only by repeated observation over many years, and by careful study of the precious few old land-use maps we have, can we understand the system of shifting cultivation that enabled farmers in the South to keep growing crops under subtropical environmental constraints.

In New England the early settlers had to cope with a set of environmental constraints that were different but equally difficult: short, cool summers, steep slopes, and infertile glacial soils. The only sizable area of truly good farmland on the entire eastern seaboard of the United States is the fertile limestone plains of southeastern Pennsylvania, and, as luck would have it, the best farmers from the Old World happened to come to the best farmland in the New.

CORN, SMALL GRAINS, AND HAY

The German-speaking farmers who settled southeastern Pennsylvania were familiar with the new four-course rotation, which had not yet been widely adopted in England. In the New World they modified it to include corn, which combines the advantages of turnips and barley. Like turnips, corn is a row crop that can be cultivated to control weeds, and it is a better feed than barley for fattening hogs and cattle. Corn became the linchpin of the Pennsylvania German four-year rotation of corn-oats-wheat-clover, and farmers carried knowledge of this rotation with them when they migrated westward.

After they had transplanted it to the Middle West, farmers streamlined the rotation by eliminating one of the small grains, either oats or wheat. Winter wheat was the usual small grain in the southern part of the region, but in the more northerly areas farmers grew oats rather than winter wheat. Winter came so early in the north that they could not plant wheat after they had harvested their corn, and winters often were so severe that the ground was frozen and the seeds in it were killed.

The new three-year rotation of corn–small grains–hay became and remained the standard rotation of the Corn Belt for more than a century (fig. 6.1). As in Europe, individual farmers modified it to fit their farms. To take two simple examples, farmers on fertile, level land might take two crops of corn before they planted oats (corn-corn-oats-hay), but those on steep land subject to erosion might protect their soil by leaving the hay crop for a second year or even longer (corn-oats-hay-hay).

AROUND THE MARGINS OF THE CORN BELT

Farmers modified the standard rotation even more in response to increasing environmental constraints around the margins of the Corn Belt. In the hills to the east and south, for example, they used the standard rotation on level areas, but much of the land was too steep for any use but pasture or woodland.

Slopes are short but steep in the recently glaciated areas north of the Corn Belt, and good soil management motivated many farmers to leave their land in hay for several years. The growing season was too short and too cool for corn to ripen dependably into grain before frost caught it, so farmers cut off the entire plant before its grain ripened, chopped it into small bits, and stored them in a silo for winter feed. Corn silage is an excellent energy source that nicely complements the protein contained in alfalfa and other leguminous hays, and the combination of corn silage and alfalfa hay is a delectable diet for dairy cows.

The plains west of the Corn Belt are so dry that corn and hay are risky, and wheat, the small grain, became the principal crop. Grain sorghum is a

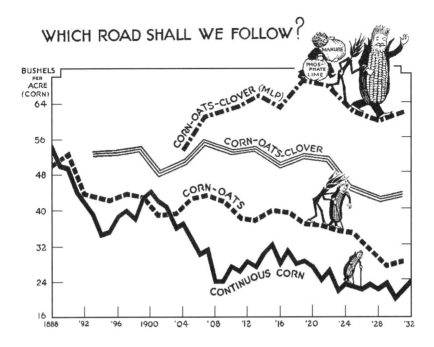

Fig. 6.1. A graphic illustration of the advantages of rotating corn with small grains and leguminous hay, which was the standard rotation in the Corn Belt for more than a century. Reproduced from Clyde E. Leighty, "Crop Rotation," in *Soils and Men*, Yearbook of Agriculture 1938 (Washington, D.C.: U.S. Department of Agriculture, 1938), 412.

Fig. 6.2. Alternating strips of cropland and fallow in a dryland wheat farming area on the Great Plains.

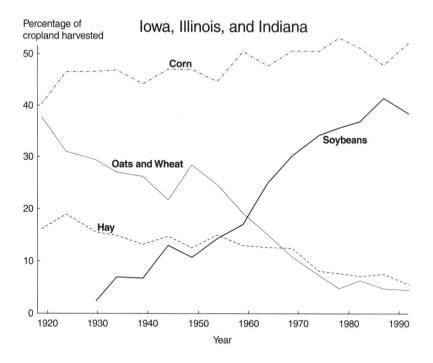

Fig. 6.3. Since World War II a new cash-grain cropping system of corn and soybeans has replaced the traditional three-year rotation of corn, small grains, and hay in the heart of the Corn Belt.

useful dryland crop on sandy soils that are too droughty for wheat. Some people erroneously assume that wheat grows best in subhumid areas. It is the dominant crop of the dryland margins only by default, because no other crop is nearly so dependable and profitable in areas that have light and undependable precipitation.

In semiarid areas even wheat is risky if farmers try to grow it every year, and they have developed a two-year dry-farming rotation of a year of wheat and a year of fallow. Any rain that falls in the fallow year is stored in the soil for the following year's wheat crop. During the fallow year the farmers cultivate the land periodically to kill any plants that might remove some of its moisture, but they keep the surface as cloddy and as trashy as possible to reduce the loss of topsoil from wind erosion. Alternating strips of green or golden wheat and rich, dark-brown fallow soil are distinctive landscape features of dry-farming areas (fig. 6.2).

The irrigated oases of the West are so far from eastern markets that many of their farmers concentrate on producing vegetables, fruits, pota-

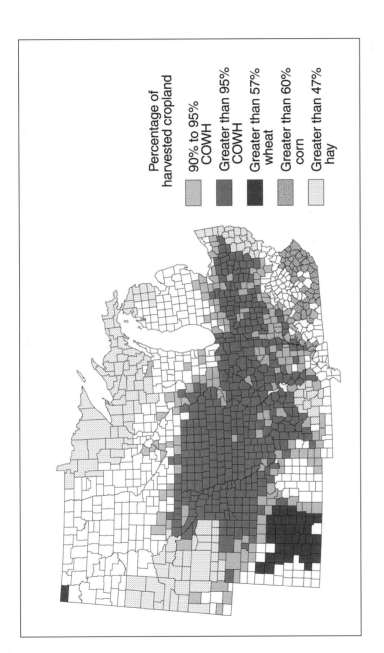

Fig. 6.4. In 1924 nine-tenths of the harvested cropland in the heart of the Corn Belt was used for the traditional rotation of corn, small grains, and hay.

Percentage of
harvested cropland

90% to 95%
COWH

Greater than 95%
COWH

Greater than 57%
wheat

Greater than 60%
corn

Greater than 47%
hay

toes, and other specialty crops that can stand the cost of long-distance transportation. These farmers know that monoculture is risky, however, and they alternate their specialty crops with feed grains, such as corn or barley, and alfalfa. The availability of large quantities of feed grain and alfalfa hay has encouraged the development of large-scale cattle-feeding operations on dry lots, so called because they have only bare ground with no plant cover.

CORN AND SOYBEANS

Improvements in agricultural technology since World War II have permitted or encouraged farmers in the Corn Belt to modify the traditional rotation. Chemical herbicides have reduced the necessity of cultivation to kill weeds, and chemical fertilizers have replaced leguminous hay crops as a source of soil nitrogen. Bigger and better machinery permits farmers to grow soybeans, which used to compete with corn for labor, and soybeans have supplanted small grains and hay in a new two-year rotation of corn and soybeans (fig. 6.3).

The standard three-year rotation of corn, small grains, and hay produced abundant feed for livestock, and farmers in the Corn Belt traditionally used most of their crops to fatten cattle and hogs. The farmers could still feed their corn to animals after they had switched to the new cropping system, but soybeans must be processed before they can be eaten, and it was easier to sell both corn and soybeans as cash crops.

There is a ready market for corn and soybeans, which are bought for feed by specialized livestock producers in the Corn Belt, in other parts of the United States, and in other parts of the world. These crops are also useful industrial raw materials. Many farmers in the Corn Belt have gotten rid of their livestock and concentrate on producing corn and soybeans for direct sale as cash crops. The traditional mixed crop-and-livestock farming system of the Corn Belt has been replaced by a new cash-grain farming system based on corn and soybeans.

In 1924 nine-tenths of the harvested cropland in the Corn Belt was used to grow the crops of the traditional rotation: corn, small grain (oats or wheat), and hay (fig. 6.4). By 1987 the changeover to the new cropping system was virtually complete in the heart of the Corn Belt, where nine-tenths of the harvested cropland was use to grow corn and soybeans (fig. 6.5).

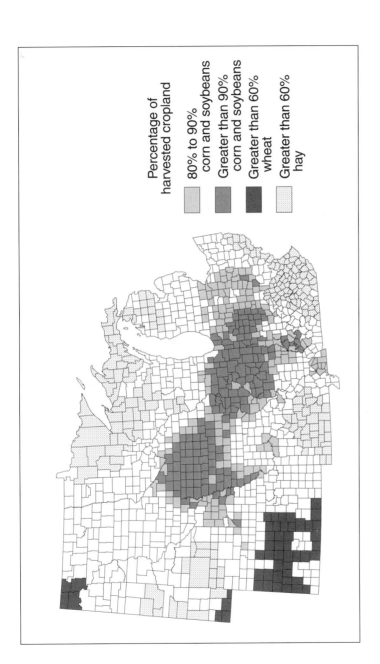

Fig. 6.5. In 1987 nine-tenths of the harvested cropland in the heart of the Corn Belt was used for the new cash-grain cropping system of corn and soybeans.

Some soybean farmers in areas with mild winters grow three crops in two years. In the first year they plant a full-season variety of soybeans, and then plant winter wheat in the stubble as soon as they have harvested the soybeans. They harvest the wheat the following summer and immediately plant a short-season variety of soybeans in the wheat stubble. They harvest these soybeans in the fall, and plow under the residue to prepare the ground for a crop the following spring. This cropping system demands hard work, excellent management, and a lot of luck. A good winter-wheat crop can compensate for reduced soybean yields, but the farmer must hope that it will also cover the additional expenses of machinery, chemicals, and labor.

Part Three

LAND DIVISION

At first blush it might seem a bit odd to describe systems of land division as structures, but few human constructs on the face of the earth are more persistent than the lines people have placed on it to divide it into countries, states, provinces, counties, townships, individual properties, even fields, and few are guarded more zealously.

Tempers would flare, for example, if Canada were to cede a few acres to the United States, or even worse, if the United States were to cede a single acre to Canada. Not many Americans have heard of the Northwest Angle, which has a year-round population of about seventy-five families, but most would be incensed if anyone had the temerity to suggest ceding this incongruous and sparsely populated bit of territory to Canada in order to straighten out a small kink in the international boundary.

State lines are equally sacrosanct. The meandering course of the Mississippi River has been straightened, but the state lines that follow its original course are just as crooked as they were when they were first drawn. Chunks of Arkansas now are east of the river, and parts of Mississippi are west of it, but no one has even dared to suggest rationalizing the boundary.

Farther downstream, near Lake Providence, Louisiana, is Stack Island, a two-thousand-acre seven-mile-long sandbar, which was once an island on the Mississippi side of the river but is now firmly attached to Louisiana. It is used mainly for deer hunting, and is actually underwater for part of the year, but twice in four years the two states have carried all the way to the Supreme Court their squabble about which state owns it.

In 1792, when Kentucky was admitted to the Union, the north bank of the Ohio River was identified as the northern boundary of the new state,

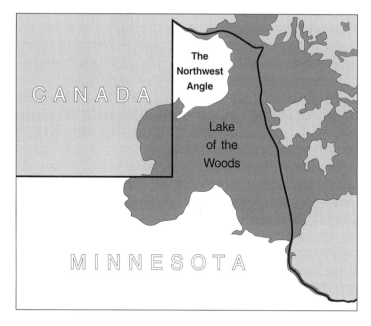

The Northwest Angle, which cannot be reached overland from the United States without going through Canada, has a year-round population of only about seventy-five people. Ceding it to Canada would remove a kink from the international boundary and incense hordes of people who presently are not even aware of its existence.

and even today residents of Ohio, Indiana, and Illinois must buy a Kentucky fishing license before they can legally drop a line in the river, even if they are sitting on their own front porches.

Municipalities, too, are fiercely protective of their boundary lines. In August 1994 the good citizens of East Williston, Long Island, were outraged to discover that two streets in the tiny affluent village had been transferred to another Nassau County legislative district in order to create a safer seat for a politician who lived on one of them.

Dividing up the land and establishing rights of ownership to it are essential first steps in the human occupancy of any area. The way in which the land is divided shows up in the boundaries of political units and their subdivisions, whether they are as large as a state or nation or as small as a farm or even a single residential plot in a suburb. It is closely related to the taxes people have to pay, to the amount of litigation over titles with which they are afflicted, and even to their attitudes toward the environment, their ability to find their way around, and their comfort. In many areas the system of land division and land survey is most immediately obvious in the patterns of the local roads, because insofar as possible they are designed to avoid cutting across individual properties.

A concern about the division of land, and about the right to use it, appears to have arisen along with the development of agriculture. Primitive hunting, fishing, and collecting groups had concepts of territoriality and

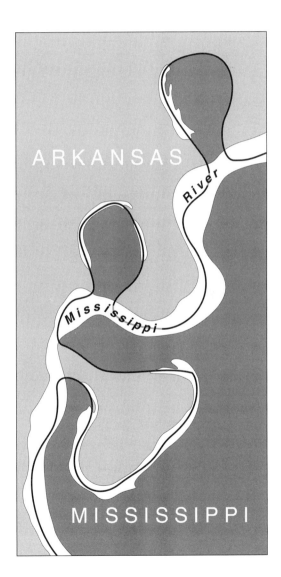

The state line between Arkansas and Mississippi follows an old course of the Mississippi River.

land division that seemed vague to Europeans. These groups were familiar with the hunting grounds and collecting areas that they and their forebears had traditionally used, and they might even be moved to violence in defense of these areas against intruders, but their notions about rights to land generally were preagricultural.

Farmers, however, feel that they have a right to the products of the land after they have tilled the soil, planted the seed, and tended the crop. They demand that others respect their rights to their own plots of ground and what they can produce from them, but in return they are prepared to accord equal respect to the rights of others to their own plots.

The right to use land may be held by an individual or it may be held in common by all members of a tribe, community, or society. This right may be exercised by the owner or it may be extended to another user, commonly in return for some consideration, such as the payment of rent, whether in cash or in kind, or the performance of specified services. In order to minimize conflict between owners it is customary to demarcate the limits of any right to land by drawing a boundary around the land to which that right pertains.

The overall system (or lack thereof) by which such boundaries are drawn is a system of land division. Systems of land division fall into two broad categories. The land may be divided before it is occupied, as happened in much of the United States, or it may be divided coincident with or even subsequent to settlement, as was common in Europe. The boundary lines that are drawn before settlement commonly are geometric, often straight lines, because people draw straight lines across areas about which they know little. The boundary lines that have evolved after settlement are more likely to be irregular because they follow natural features, such as ridgelines and rivers, that are easily identifiable on the ground. The boundaries of the Canadian provinces are mostly straight lines, the state map of the United States has good examples of both kinds of boundary lines, and it is well-nigh impossible to find straight lines on the political map of Europe.

Systems of land division in the United States and Canada were strongly influenced by the ideas that people brought with them from Europe, and especially from the British Isles, because people from Europe, and especially people from Britain, were responsible for most of the land-survey systems that were imposed on the face of North America.

7 *Land Division in Britain*

The British Isles are a fine museum in which to see systems of land division established coincident with or subsequent to settlement. British historical geographers enjoy describing their island as a palimpsest. A palimpsest was a piece of specially prepared sheepskin or goatskin on which medieval scribes wrote time and again. Each time they wrote, they scraped off the earlier writing, but nonetheless faint traces still showed through, just as faint traces of the past still are visible in the present landscape if one knows how to look for them.

Not only does Britain have specimens of most of the major kinds of land division that have been used in western Europe, but it also has a distinctive Atlantic Fringe, which few parts of the Continent can match. Furthermore, most of the early settlers of North America came from the British Isles, and part of the cultural baggage they brought with them was British ideas about how land had been and could be divided and tenured.

It is not easy for an American to comprehend the vast timespan of British history. The Roman occupation lasted four hundred years, the Normans arrived six hundred years after the last Roman legion had departed, it was four hundred years after the Norman Conquest that Columbus set foot in the New World, and just think of how the New World has changed in the five hundred years since Columbus. The Old World has changed just as much, and it is reasonable to assume that earlier eras had transformations of comparable magnitude.

It is equally reasonable to assume that rural Britain had as much geographical diversity at any time in the past as it has today. One might argue that every settlement and perhaps even every field has a unique history, but it is useful to step back and try to see the forest while ignoring the trees (not to mention the branches, and even some of the twigs, which have so fascinated some scholars). Any brief broad-brush sketch of necessity must oversimplify a complex reality, but it can identify some of the systems of land division and land tenure that still are evident in the contemporary land-scape of rural Britain, and ponder their possible impact on the American rural scene.

Traditional scholars have described British prehistory as a series of invasions by technologically superior groups from the Continent—the Celts, the Beaker folk, the Belgae. Each new group grabbed the best land and pushed the native people farther and farther into the less tractable, rain-sodden hills of the west and north. Such an interpretation certainly enables the English to feel superior to the Welsh and the Scots. It seems unnecessary to postulate any such invasions, however, because the local people could have acquired new technologies from the Continent just as easily by trade and other contacts as by invasion.

The recent development of sophisticated paleoecological techniques, such as pollen and soil analysis and the study of prehistoric snails and insects, has resulted in a better understanding of environmental changes in times gone by, and some scholars now interpret the history of rural Britain in terms of recurring cycles of environmental crisis rather than invasions. For centuries, they say, the population increased until it exceeded the abil-ity of the land to support it at a given level of technology. Population pres-sure then led to famine and disease, warfare, overexploitation of the land, and soil depletion and erosion. Population declined, settlements con-tracted, and land was abandoned. Trees slowly recolonized the abandoned land and began to heal it until the availability of a new technology that could use the land more effectively encouraged a renewed attack on the forests and launched yet another Malthusian cycle.

The earliest recognizable human inhabitants of Britain were small kin groups of Paleolithic hunters and gatherers whose lifestyle resembled that of the nomadic Indians on the American Great Plains. Each of their base camps, clusters of crude wooden huts, was six miles or so from the nearest neighboring cluster, but for much of the year these people camped in skin tents while they followed the seasonal migrations of the herds of wild cat-tle and deer that met most of their needs. The men hunted while the women and children collected everything their experience had taught them was fit to eat. The groups were not free to wander about as they wished, because each one had its own hunting territory of thirty-six to fifty square miles, and it knew and respected the territorial rights of neighboring groups. Its members traded with the members of other groups, especially for the pre-cious flint they needed to make their stone tools.

The Neolithic revolution, which ended the Paleolithic era, was one of the most important epochs in human history, because it was then that people learned to domesticate animals instead of hunting them in the wild and to cultivate plants instead of depending on what nature happened to provide. Despite our vaunted modern knowledge of science and technology, we have domesticated no new plants or animals since the Neolithic revolution, and we still depend on those that were domesticated by our illiterate primitive ancestors.

The innovations of the Neolithic revolution originated in the oases of the Middle East. They diffused slowly across Europe and came late to Britain. The Neolithic people vigorously attacked the forests with their superb stone axes and burned them to make the land available for cropping. They were good farmers who grew wheat, barley, and other cereals, and they had herds of cattle, sheep, goats, and hogs. They cultivated small plots of half an acre or so with light wooden plows. These plows dug straight furrows and cut the sod, but they had no moldboards to turn it, so the farmers had to cross-plow their fields at right angles to secure the necessary tilth. Cross-plowing produced small, squarish fields that show up clearly on aerial photographs. Most of those that can still be identified are in marginal farming areas, because subsequent plowing has obliterated all traces of them in the better areas where they once were common.

The Neolithic people lived in isolated clusters of thatched wooden huts. They rarely occupied a house for more than twenty-five years or so. They built a new one close by when the timbers in the old one started to rot, but they cultivated the same "farms" for centuries. Their farms were parts of larger tribal territories bounded by continuous banks or ditches that extended for miles across the countryside and enclosed land units of hundreds of acres; this suggests that they had chieftains who ruled considerable areas. Their high degree of political, economic, and social organization is also suggested by their great religious monuments, such as Stonehenge, by their numerous and impressive burial mounds, and by the evidence that they carried on a complex long-distance trade in axes and pottery, and probably also in animals, hides, salt, and other goods.

There is evidence for a widespread environmental crisis toward the end of the Neolithic period. Settlements were shrinking, farmland was being abandoned, and the wooded area was increasing. This trend was reversed by the introduction of metal tools, first bronze, then iron, which encouraged a renewed attack on the forest and the formation of many new settlements. During the Bronze and Iron Ages England was settled as densely and farmed as intensively as during the height of the Middle Ages.

The farms were parts of larger territorial units that were ruled by warrior chieftains from the hilltop fortresses of all sizes and shapes that are familiar to anyone who has used a British topographic map. Some hill forts were inhabited, but many were merely places of refuge when trouble threatened. The proliferation of metal spears, swords, and other weapons in archaeological finds of the late Bronze Age suggests that the growth of population had forced people to fight to protect what they had or to seize what they needed, and many farmhouses were enclosed by banks of earth surrounded by ditches, with stout wooden palisades on both sides of the ditch.

The establishment of well-defined and persistent territorial divisions was one of the most important developments of the Bronze Age. These divisions have persisted because they have formed the basic units of an agricultural system that changed only slightly for two thousand years or more, and it was extremely difficult to alter them once they had been established.

Each unit had a nucleus of cropland on the better soils, with meadows for hay along the streams and in the wetter parts, and woodlands or wastelands of bog, heath, and moor in the poorer marginal areas. The unit could support a group of farmers and produce enough surplus to maintain an overlord. Every farmer was the vassal of an overlord to whom he had to give some set share of what he produced. Presumably at one time each unit had a single overlord. Overlordships could be divided by inheritance, however, so the different farmers in a unit might have different overlords even as the practical necessities of farming kept them working together in the unit. Sometimes they lived on isolated farmsteads, sometimes in small clusters, but their homes remained within the stable boundaries of the unit.

Some of these territorial units were already old by Roman times, and classical authors have identified by name some of the tribal territorial units that existed in southern England at the end of the Iron Age. The Romans and later the Saxons and Normans merely seized control of the existing units, which continued to operate much as before. The medieval church used them as its ecclesiastical parishes, and they persist even today as civil parishes. It is awesome to look at a modern boundary line and realize that some of the stones that mark its course might have been placed there more than four thousand years ago.

The Roman conquest actually had little effect on the organization of British agriculture because only a few administrators followed the Roman armies to Britain. The local people might have adopted Roman manners and dress and building styles, but the local chieftains continued to rule their people as overlords under the aegis of the Roman Empire. British farmers prospered under the Pax Romana, and a flourishing export trade in grain and wool enabled enterprising overlords to organize their territorial units into efficient estates. The administrative centers of these estates were villas with solidly built farmhouses, barns, and a range of other outbuildings set around courtyards, but the field systems and territorial units remained essentially unchanged.

The Roman era was one of the periods in English history—the sixteenth and seventeenth centuries were another—when prosperous landlords could afford to erect great country houses to advertise their newfound success and rising status. The most significant Roman contribution to the contemporary English rural landscape was not these villas, however, because they have all been destroyed. Far more significant for later generations was the construction of more than five thousand miles of roads, many running straight as an arrow across the countryside. The lines followed by many of these roads are still in use today, but more important, the roads provided ready avenues by which the later Saxon invaders could penetrate wild and unsettled country more readily than if they had had to hack their own way through it.

THE CELTIC FRINGE

Before considering the Saxon invasion, however, it is appropriate to examine systems of land division and land tenure in the Celtic hills of the west and north, from which many people migrated to North America. These rugged uplands, swept by wind and lashed by rain, are uncharitable to people and their animals. The tough old Paleozoic rocks, stumps of ancient mountains, have been scraped bare by glacial ice and plastered by blankets of bog and acidic, infertile, peaty soils. The small pockets of even moderately good land are few and far between. This forbidding land was the refuge and the home of the Celtic people.

The Celts practiced an infield-outfield farming system that was in marvelous harmony with the difficult environment. For the most part it has disappeared in modern times, but it still persists in a few remote areas. The infield was the best patch of ground available. It was enclosed by a stone wall or high earthen bank, was cultivated continuously, and received all the farm manure. The standard crop was oats, which nourished both animals and people. The outfield, of lesser fertility, was essentially pasture, but it was cultivated under a system of "long ley." The farmers enclosed a patch of the outfield with a temporary fold the summer before it was scheduled for cultivation, and penned cattle on it to enrich the soil with their droppings. Then they took a crop or two of oats or potatoes from it before they returned the land to grass until the time came to cultivate it once again.

Each part of the outfield was cultivated once every seven to ten years. Regular cultivation was necessary to renew the grazing quality of the pastures, which deteriorates fairly quickly under the excessive moisture conditions of the Celtic lands. Even today some farmers in Scotland talk about the necessity of "taking the plow around the farm" in order to maintain the quality of their pastures. The infield-outfield system had accustomed farmers in Scotland and Ireland to the idea of land rotation and the long fallowing of exhausted land well before the first Scotch-Irish immigrants came to the United States, and this part of the cultural baggage of the Scotch-Irish settlers might have helped initiate the brush-fallowing system of land rotation they practiced in Appalachia.

In the Celtic areas the uphill margin of the outfield, which was the upper limit of cultivated land, was enclosed by a stone wall, or head-dyke. The farmers pastured their sheep and cattle on the improved grasslands below the head-dyke, as well as on the rough grazings of heather and coarser grasses on the moorlands above it. In addition, they took some animals to distant upland pastures during the summer cropping season, which in Ireland traditionally began on Saint Patrick's Day and ended on Halloween. The young folk, and sometimes entire families, left their winter homes and took their livestock and even some of their furniture to the summer pastures. There each family lived in a hut known as a *booley* in Ireland, a *hafod* in Wales, a *shieling* in Scotland, and a *saeter* in Norway. Both the departure in spring and the return to the winter home in fall provided excuses for great ceremonies and festivities, some of which are still celebrated today, even though many contemporary celebrants have not the foggiest idea of their original significance.

The most common type of winter home was a "long house," a low, one-story, rectangular cottage that sheltered the family in one end and cattle in the other (fig. 7.1). Long houses were twelve to fifteen feet wide and of varying lengths, and might be built of any material that was locally available—stone, timber, clay, wattle, even sod. Each cottage sat on a slope at right angles to the contour, and the top end, with chimney, fireplace, and living quarters, was built into the excavated hillside for warmth.

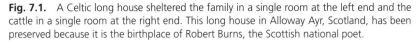

Fig. 7.1.   A Celtic long house sheltered the family in a single room at the left end and the cattle in a single room at the right end. This long house in Alloway Ayr, Scotland, has been preserved because it is the birthplace of Robert Burns, the Scottish national poet.

The cattle were kept in the lower end of the long house, with the slope permitting refuse to drain away. Food for people and feed for their beasts were often in short supply during the hard winter months, and it was not unusual for a farmer to drain a bowl of blood from one of the live animals and mix it with oatmeal to make a nourishing blood pudding for the family table. Come spring the cattle might have to be carried out to pasture because they had been bled so often they were too weak to walk.

The contemporary farmsteads of the Celtic fringe are dispersed on their own plots of land, but over the centuries these areas apparently have experienced more than one cycle of settlement nucleation and dispersal. The growth of the population, combined with inheritance laws requiring that a man's property be divided equally among his sons, could have converted the isolated farmhouse of a father into a cluster of farmhouses inhabited by his sons and grandsons. Ireland, for example, once had many small, irregular clusters (known as *clachans*) of ten to twenty farmhouses in which nearly everyone had the same last name and presumably a common ancestor. Nicknames often were necessary in clachans where everyone had the same surname. On the other hand, a decline in population, such as resulted from the Great Famine of 1845 in Ireland, could have enabled a single survivor to consolidate all the land back into a single farm.

The patterns of land tenure of many clachans suggest that all the farmers were descended from a common ancestor, and were farming as a group the land that he had once farmed himself. No farmer held a solid block of land, but all held plots scattered across the infield, as though it had been divided to give each person a share of each kind of land, then subdivided and divided once again, often resulting in impossibly small plots. For example, in one clachan twenty-nine farmers had 422 plots of land, in another one person had 32 different plots, and in yet another no fewer than twenty-six people held shares in a single half-acre plot. Much of the good land was wasted in boundaries, and many farmers had to trudge hundreds of miles each year just going from one plot to another. The problems of such minuscule subdivision are obvious, but land consolidation is a difficult and tedious process because every owner fears having to exchange better fields for poorer fields.

OPEN FIELDS

"The contrast between the villages of the plain with their strip fields and the scattered farms of the Atlantic coast lands with their hedged fields is striking," said Estyn Evans (in "The Ecology of Peasant Life in Western Europe"), "and the attitudes of the folk who live and work in them are correspondingly different. The English village has lost many of its old functions and has changed its social structure, but there can be little doubt that it bred a disciplined tradition and a respect for law and order which, to the Irishman, appear to be the mark of simple minds."

Few contemporary English villages can be traced back before around A.D. 450, when the Romans left and the Saxon invasions began, yet the names of most of them are recorded in the Domesday Book of 1086, which was compiled twenty years after the Norman Conquest. Historians have invoked these facts to support their belief that the traditional English countryside of nucleated villages and open fields was created during the Saxon period, after the departure of the Romans and before the arrival of the Normans. The Saxons, the historians have said, invaded an area that was still largely wilderness and transformed it into a settled agricultural land. They killed or drove out the natives, established nucleated villages, cleared the primeval forest, and laid out large open fields that were divided into strips but cultivated in common.

Recently archaeologists have begun to challenge the conventional wisdom of historians. They argue that the Saxon invasions were merely a hostile political takeover of a society that had already started to disintegrate as Roman rule grew weaker. New Saxon overlords seized control of existing agricultural units from native chieftains, but the Saxons did not introduce new farming methods or field systems. The ordinary people continued to work the land as their forebears had done for hundreds of years. They continued to live on isolated farmsteads, not in villages, and over the centuries they continued to shift their farmsteads from place to place within the territorial units that dated back to the Bronze Age.

Historians and archaeologists agree that the Saxons cleared extensive areas of woodland, especially during the latter part of their era, but it was secondary growth that had recolonized farmland after the Romans had departed, not primeval forest. During the Roman era the countryside was densely populated and sparsely wooded, but the population declined sharply during the chaotic period of political and economic instability after Roman rule collapsed and tribal warfare resumed, and much farmland was abandoned to woodland in the early part of the Saxon era, then cleared once again in the latter part.

One nagging question still remains: Why do the names of villages appear in the Domesday Book if they were not even founded until later? Archaeologists reply that the Domesday Book was a record of landholdings, not of settlements. The names in it are the names of longstanding territorial units, some of which date back to the Bronze Age, and not the names of villages, which were founded later. The villages, when eventually they were founded, took the names of their territorial units rather than vice versa.

Then when did the traditional English landscape of open fields and nucleated villages originate? The best current answer seems to be that it was deliberately created in the first few generations after the Norman Conquest of A.D. 1066, when the new Norman overlords, who had replaced the former Saxon chieftains, began to rationalize, reorganize, and modernize their new estates. They left no written record of this reorganization, but archaeological evidence indicates that it occurred around A.D. 1100 or shortly thereafter.

The new Norman rulers each reorganized the cropland of their estates into three large unfenced fields to fit the standard three-year rotation of food grain, feed grain, and fallow. They divided each large field into many smaller strips that were held by individual peasants, but the peasants farmed the entire field in common because few were wealthy enough to own a heavy plow and the eight oxen needed to pull it; one might contribute the plow, another a team of oxen, and so forth.

The standard strip was an acre, or the amount of land a team of oxen could plow in one day (fig. 7.2). It was a "furrow long" (furlong), or the distance the team could pull the plow before it had to stop and rest. The length and width of strips varied with the nature of the soil and the lie of the land, but an acre became standardized as a strip 220 yards long and 22 yards wide. These dimensions remain with us; racetracks still measure distances in furlongs of one-eighth of a mile, and 22 yards (or 66 feet) is exactly the length of a cricket pitch, and just a little more than the distance from the pitcher's mound to home plate in baseball. Imagine high-spirited peasant lads hurling clods of earth at each other back and forth across a strip.

The peasants turned each slice of soil cut by the plow toward the center of the strip when they plowed it. Over time this process built up a slight ridge down the center of the strip, and left low furrows between each strip and those on either side of it. These furrows not only marked the boundaries between strips, but they also facilitated drainage in a moist climate. Over the years this process built up a corduroy, or "ridge and furrow," surface that can still be seen in many fields in the English Midlands. Neither the strips nor the fields were enclosed by fences of any kind, thus creating the historic landscape of open fields that was common throughout medieval Europe, not just England.

Each peasant had the right to keep a certain number of livestock, which were tended by old people and children to keep them out of the open fields. At night the cattle and sheep were kept in their stalls, or in the fields in temporary enclosures called folds; during the day they were turned onto the common wastelands, the fallow field, the meadows after they had been mown for hay, or the stubble in the barley field after the grain had been harvested in the fall. Grazing rights were so precious that they were strictly regulated, and a vast corpus of common law controlled the use of common land.

The lack of fences necessitated oxherds, cowherds, and shepherds, such as Little Boy Blue in the nursery rhyme, who failed to keep the sheep out of the meadow and the cows out of the wheat (which the English call corn). Swine ran loose in the woods in care of a swineherd; in the medieval ballads, when a traveler encountered a wild man in the forest, he was almost invariably there tending swine.

Fig. 7.2. Cultivating a strip in a vast open field with oxen.

The land of each territorial unit originally was separated from the land of other units by empty wastelands of forest, bog, heath, and moorland, and thus it required no sharp delineation. As the population increased, however, these wastelands were reclaimed and brought under cultivation to meet the growing demand for food, and it became ever more important to demarcate the boundaries sharply to avoid territorial conflicts. Boundaries were defined by prominent features such as roads, stones, trees, and even, in one instance, "the beech tree wherefrom the thief was hung." Young men were expected to learn the boundaries, or "marches," which one of the elders showed them in an annual perambulation, and in some areas the ceremony of "riding the marches" is still celebrated as a festive summer event.

VILLAGE, MANOR, AND PARISH

Under the new system every peasant had to have an equal number of strips in each of the three fields, and everyone's strips were scattered so far and wide that the only sensible place for them to live was at a central location. The Norman rulers razed the old isolated farmsteads of the Saxons and moved the people into planned villages with rows of cottages on identical plots facing straight streets or geometric village greens. Back lanes gave access to the enclosed crofts, paddocks, or garden plots behind each cottage. Cottages were abandoned in later years, when the population declined, and new cottages were built at different sites when it grew, so most villages have been greatly modified in the eight hundred years since they were first laid out. Often only careful archaeological investigation can identify their original plans.

The Normans imposed feudal rule on the villages and the farmland associated with them. The feudal system was the standard form of political organization in medieval Europe at a time when the central government was too weak to maintain effective law and order and when the economy was almost purely agricultural, based on barter and the exchange of services rather than on money. Every man had an overlord to whom he rendered certain services and from whom he expected leadership, protection, and justice. The duties of government passed into the hands of a fighting aristocracy of hereditary overlords who protected the peasants who tilled the soil for them.

Under the feudal system no one actually owned any land; they merely held it "in feu" in return for service to an overlord. In theory, the king was God's chief tenant for his entire kingdom by divine right of birth. He parceled it out among his nobles in return for the services, mainly military, that he expected them to render to him. He assigned principalities to princes, duchies to dukes, marches to marquesses, earldoms to earls, counties to counts, and baronies to barons. Each of these worthies, in turn, divided his land into smaller units that he assigned, once again, in return for military services.

At the bottom of the feudal hierarchy was the smallest unit of land, the manor, which was just large enough to provide one gentleman or knight, the lord of the manor, with a livelihood for himself and his family, plus the horse, armor, and weapons he needed to render military service to his lord and to protect the peasants who worked his land. Many, perhaps most, manors were the same territorial units that had been established during the Bronze Age and adopted as ecclesiastical parishes by the early Christian church.

At the village or manorial level the peasants maintained by their labor the lord of the manor, to whom they owed a set share of what they produced, for whom they were obligated to do such jobs as harvesting, woodcutting, and road repair, and under whose command they fought. Each

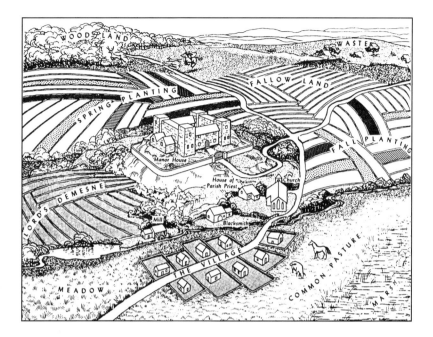

Fig. 7.3. A stylized diagram of a medieval manor with three open fields and a central village. Reproduced by permission of the publisher from Wallace K. Ferguson and Geoffrey Bruun, *A Survey of European Civilization*, 4th ed. (Boston: Houghton Mifflin, 1969), 155.

peasant had to work a certain number of days on the demesne land that belonged to the lord and on the glebe land that belonged to the church. Virtually every activity in the village, every right and every duty (grazing rights, labor services that had to be performed, payments in kind, and so forth), came to be defined by a vast body of common law that was "the custom of the manor as interpreted in the lord's court."

No two manors were exactly alike, but in its simplest form a manor consisted of one village and one system of three open fields that coincided with an ecclesiastical parish and constituted the estate of a single landlord or squire (fig. 7.3). Its most imposing structure was the home of the lord of the manor, which might have been a fortified castle in the early days. Later many castles were replaced by more-livable mansions, or halls. Close by the hall was the parish church and the home of the parish priest.

The peasants lived in the village, which had no stores, but it did have workshops such as the smithy, where the blacksmith made nails, shoes for horses and oxen, and plow irons (fig. 7.4). Beside the stream was a water

Fig. 7.4. A modern agricultural village with strip fields in the foreground.

mill for grinding grain. The lord claimed the right to have all the peasants grind their grain at his mill, and claimed part of it as payment for this service. The lord often leased this right to a miller, and millers were crafty fellows, not above taking more than their share. Many villagers preferred to grind their grain at home with old-fashioned hand mills, or stone querns. The exasperated abbot of St. Albans, after half a century of angry disputation with his villagers, finally confiscated and broke all their quern stones, and paved the floor of his parlor with the pieces.

Although traces of the feudal past still linger on in such modern names as the principality of Wales, the duchies of Cornwall and Lancaster, the marches of Scotland and Wales, and the ubiquitous "county," which crossed the Atlantic Ocean to all fifty states except Louisiana and Alaska, the principal relic of the feudal system is the basic unit of landholding and decision making, the manor. Over the centuries, in a fuzzy fashion, the manor was superseded for administrative purposes by the ecclesiastical parish, partly because the church administered the poor laws, partly because the relationships among field systems, nucleated villages, manors, and parishes were vastly more complicated than this simple sketch has suggested.

Nevertheless, village, manor, and parish were still held together because the people of the village were expected to go, and for centuries did go, to the parish church, and because the greater part of the village and its land were owned by a single squire, who played a key role in deciding what would happen to them. The ecclesiastical parishes eventually were converted into civil parishes, which are the smallest geographic units for which agricultural and demographic statistics are available in Britain.

LAXTON

Most of the open fields were enclosed more than a century ago, when lowland England was converted into a countryside of hedgerows, but a few museum pieces remain to give some notion of what the open fields once were like. Fragments of the older system have survived in several places, but the only complete open-field village that still operates is Laxton in Nottinghamshire, some twenty miles north and a bit east of the city of Nottingham. In 1952 the British Ministry of Agriculture purchased the manor of Laxton with the intention of maintaining it as a working example of open-field farming. This system of farming is completely obsolete, however, and even though rents are low, the costs of moving animals, implements, seeds, manure, and crops are so great that the farmers can make only a modest living, and they may well decide that they do not wish to make the sacrifices demanded of human exhibits in an open-air museum.

The parish of Laxton (originally Lexington, the *tun*, or farm, of Leaxa's people) probably was occupied by Saxon farmers in the sixth or seventh century. It is mentioned in the Domesday Book, and some of the strips in the modern open fields may have been laid out more than eight hundred years ago, but the history of Laxton really begins in 1635, when a London merchant, Sir William Courten, bought the manor and commissioned a surveyor named Mark Pierce to make a detailed map showing each strip in the open fields and the name of the person who held it. The original copy of this map is preserved in the Bodleian Library at Oxford, and it is faithfully reproduced in Orwin and Orwin, *The Open Fields*.

In 1635 the strips of the demesne had already been consolidated into a solid block of land just north of the village of Laxton (fig. 7.5). The inner portions of the four great fields (West, Mill, East, and South) were laid out in bundles of strips and cultivated in a three-year rotation; West Field (318 acres) and East Field (134 acres) were treated as a single unit of roughly the same size as Mill Field (433 acres) and South Field (428 acres).

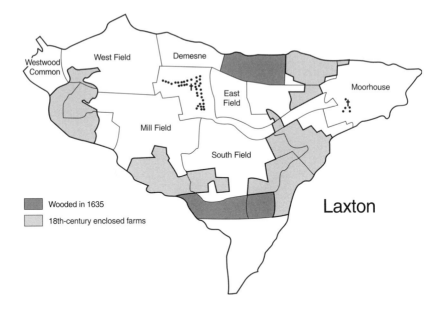

Fig. 7.5. The village of Laxton, northeast of Nottingham, England, where the traditional open-field system is preserved as a museum. Compiled from maps in C. S. Orwin and C. S. Orwin, *The Open Fields*, 3d ed. (London: Oxford University Press, 1967), and J. D. Chambers, *Laxton: The Last English Open Field Village* (London: H.M.S.O., 1964).

The more distant areas, which were reclaimed later than the open fields, consisted of solidly enclosed blocks, or "closes," except Westwood Common, which provided all the villagers with grazing land for their livestock when crops were in the ground. The Long Meadow, which was divided into narrow individual strips from which the villagers could cut their hay, occupied the wet low-lying land along the brook that flows past the southern end of Laxton village toward the hamlet of Laxton Moorhouse, at the eastern end of the manor.

The daughter village of Laxton Moorhouse was a miniature version of the mother village of Laxton, with its own open fields, its own meadowland, and its own common land. It even had its own chapel, which was what the English called a place of worship subordinate to a parish church, in this instance the parish church in Laxton village. Laxton Moorhouse was established in the thirteenth century, when the growing population of Laxton required more land than could conveniently be farmed from the village itself, so part of the community was hived off to develop its own separate economic life in a remote corner of the manor. The distant southern tip of the manor was made into a separate submanor at about the same time. Much later, in the eighteenth century, some of the closes and wasteland at the margins of the manor were consolidated into solid blocks, withdrawn from the open fields, and made into five separate farms with their own new farmhouses and other buildings.

ENCLOSURE AND THE AGRICULTURAL REVOLUTION

The medieval manor produced all it needed, but needed all it produced; the manorial system was essentially a closed, domestic, subcommercial economy. There was little surplus for trade, and the exchange of goods consisted largely of barter at the local level. Landownership conferred prestige, but it had little monetary value at a time when the rendering of services, rather than money, was the principal medium of exchange. But as the economy gradually became more commercial and as money began to circulate more freely, landlords and their tenants were increasingly tempted to commute customary services into payments of money; the prosperous tenant might have preferred to pay rent instead of having to work on the lord's land, and the landlord might have been happy to convert customary services into cash income. The desires of landlord and tenant often failed to coincide, however, and the long history of the demise of feudalism is the sorry tale of a series of wrangles between landlords who insisted on customary services, or demanded money instead of services, and tenants who insisted on commuting these services, or demanded their right to perform them instead of paying rent.

The two most dramatic episodes in the decline of feudalism and the open-field system in England occurred in times of rural depopulation and rapid agricultural change. Both were periods of fairly widespread enclosure of open fields by hedges, stone walls, and other kinds of fences. The first wave of enclosures is associated with the Black Death, which broke out in England in 1348 and wiped out a quarter to a third of the total population. The vastly reduced rural labor force was inadequate to maintain the traditional open-field cropping system at just the time when landlords were beginning to realize the profits they could make by converting the open fields to pastures for sheep and cattle. In many villages the landlord evicted the dwindling band of surviving villagers, demolished their mud-walled hovels, and enclosed the land for grazing.

The first wave of enclosures excited fierce antagonisms against the landlords and great public sympathy for the evicted villagers, and some rather ineffectual laws were passed that forbade enclosure. Public policy changed, however, around 1660, in response to the innovations that triggered the Agricultural Revolution, and the second wave of enclosures began to gather momentum. It peaked between 1750 and 1850, when private acts of Parliament approved the enclosure of some four and a half million acres of land in nearly three thousand parishes, mainly on the Midland Plain. This second wave of enclosures coincided with the beginnings of the Industrial Revolution, for which it provided food and labor; many farmworkers pushed off the land by enclosure were pulled into the burgeoning urban industrial centers.

The second wave of enclosures was vigorously promoted by progressive landlords because the restrictions of the open-field system severely limited their ability to innovate. In the open fields, for example, it was impossible to grow a crop such as clover, which would still be in the ground after the traditional harvest date, because the villagers forcefully defended their right of common grazing on the stubble after that date. (The early settlers who brought this attitude with them to Massachusetts safeguarded their common lands so carefully, and in perpetuity, that reportedly it is still quite legal to graze a cow on Boston Common.)

The Agricultural Revolution of the eighteenth century was an upward spiral involving the enclosure of the open fields, the introduction of new crops, a new rotation system, and larger and better livestock. The new rotation of wheat, roots, barley, and clover produced more and better feed, which enabled the farmers to keep more and better livestock, and the animals no longer had to be half-starved during the winter months. On the newly enclosed pastures, farmers could improve the stock by controlled breeding practices that would have been impossible on the old open fields. The larger, more numerous, and better-fed animals produced larger quantities of manure, which was returned to the fields to produce higher yields of grain. Some crops were eaten directly in the fields by flocks of sheep that were penned inside temporary fences, or folds, which not only fertilized the soil and helped maintain its fertility but also cut down on harvesting costs.

Not all the land that was enclosed was cultivated. Although many former strips of ridge and furrow have long since disappeared under the plow, in some areas enclosure removed the plow as well as the plowman, and since enclosure the land has been used mainly to pasture cattle and sheep.

Over large parts of the English Midlands, for example, the old pattern of ridge and furrow is still clearly visible, especially when the sun hangs low in the sky and the shadows lie long upon the land.

THE NEW RURAL LANDSCAPE

Although traces of the older open-field system can still be seen in many areas, over much of lowland England the enclosure movement and the Agricultural Revolution produced a new rural landscape, a patchwork quilt of small fields stitched together by hedgerows (fig. 7.6). The newly enclosed field was surrounded by a ditch, which marked its boundaries and facilitated drainage. Soil from the ditch was thrown up into a mound on the inner side, and this mound was planted with cuttings of hawthorn or other plants that would grow to produce a quickset (live) hedge. Ash or elm trees might be planted along the hedgerow to provide shelter, wood for domestic purposes, and aesthetic satisfaction. A good quickset hedge can last forever, because it can be *laid* or *pleached* to renew its stocktightness when it starts to get old and stemmy (fig. 7.7). The pleacher cuts about halfway

Fig. 7.6.   An enclosed hedgerow landscape.

Fig. 7.7. A laid quickset hedge.

through each woody stem near its base, bends the stem over into a horizontal position, weaves it around four-foot staves that have been driven firmly into the ground, and weaves together the tops of the staves with twisted sprays of blackberry or willow. The top side of the laid stem sends up new shoots, the bottom side puts down adventitious roots, and soon the hedge is as good as it ever was. Furthermore, a laid hedge is better than a stone wall for a field of foxhunters hallooing across the countryside, because it does less damage if someone happens to crash into it.

On many manors the demesne was the first land that was enclosed, especially if the lord of the manor had been clever enough to make judicious trades and thus consolidate his open-field strips in a contiguous block of land. When customary services began to be commuted, and the lord could no longer depend on the peasants to cultivate the demesne, he might have had to hire wage labor, or he might have rented the entire farm to an ambitious tenant. Today the land of the former demesne is often called the "home farm" of the village, parish, or manor, and the landlord or the landlord's representative manages it directly.

Awkward problems of farm layout were created when the land of a manor was enclosed and divided into individual farm units consisting of solid blocks. The farm buildings were all in the village, so the new farms had to be pie-shaped, with their points in the village where their buildings were (fig. 7.8). A pie-shaped farm is difficult to operate, however, and eventually the farmers alleviated this difficulty either by building a new field barn at the far end of the farm or by building a completely new farmstead

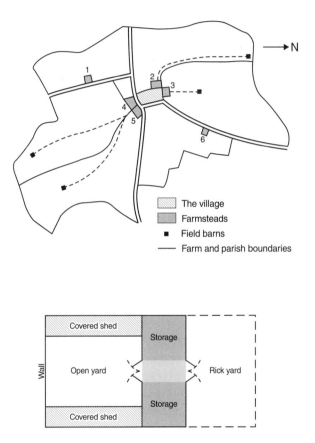

Fig. 7.8. The layout of farms in an English village (*top*) and the ground plan of a field barn (*bottom*). Before enclosure all farmsteads were in the village, and each enclosed farm was made pie-shaped to provide access to its farmstead. Some farms have new farmsteads at more-convenient sites, but others make do with a field barn at the back of the farm. The field barn saves the labor of having to haul the grain all the way in to the farmstead, and the onerous chore of having to haul the manure all the way back to the fields. A field barn is a standard English three-bay barn, with a rickyard in front where unthreshed sheaves of grain were stacked in ricks, a rectangular barn with large double doors in either side opening onto a central threshing floor, and an open stockyard behind, where cattle could be fed for production of meat and manure.

Fig. 7.9. Harvesting potatoes in southwestern Scotland. Farmers are prodigal of labor in Europe, where labor is cheap relative to land, unlike North America, where land is cheap relative to labor.

near the center of the farm when the old farmstead in the village began to deteriorate and require replacement.

Many of the people who lived in the village were farm laborers who owned no land. Unlike in North America, where land is cheap and labor is dear, in Europe land is so precious that farmers are prodigal of labor, which is relatively cheap, and six or eight workers might be employed on a farm that could support no more than a father and son in the Middle West (fig. 7.9). In the United States the first adult one meets on a farm is probably the farmer, but in England, where half a dozen workers are found in a field, the farmer is the one who still has a jacket on; the workers have taken their jackets off and hung them on the hedge. The farmworkers who live in the village form the lowest and largest group in a three-tiered rural social structure. Well above them is their employer, the farmer who manages the land, but the farmer is a tenant of the landlord who owns it.

Contrary to the beliefs of some rural sociologists, to whom any form of farm tenancy apparently is anathema, most tenant farmers in lowland England are quite content with their status, and many of them would not even consider buying the land they are farming unless forced to do so.

Before 1883 English tenant farmers might have been at the mercy of land-lords, but a series of laws passed since then has given the farmers ever-increasing freedom and security of tenure, and today they can be removed from their tenancy only if convicted of bad farming practices before juries of their fellow farmers, who are understandably reluctant to pass judgment on their own. Rents, which often have been set some years ago, are unrealistically low, and on many estates the rents do not even cover the cost of taxes, upkeep, and management, much less provide any return on the capital invested in the land.

Landlords have the prestige of landownership, and they may be reluctant to rupture their relationships with their tenants by haggling over rents. When a tenancy falls vacant, however, a landlord may prefer to sell the land rather than taking on a new tenant, and often a farm must be sold when a landlord dies and heavy death taxes are levied against the estate. The sitting tenants commonly are given first refusal, and usually buy the land, but they may have to borrow part of the purchase price. Purchase of the land absorbs some of their working capital, and if too much money is tied up in land and buildings they may be seriously short of the money needed for the operation and maintenance of the farms. Many farmers in England prefer to be tenants, and they have no desire to become landowners.

8 *Land Division in America*

The early European colonists who settled the New World came from areas with strong feudal traditions, where land was the principal source of wealth, prestige, and power. Land was so precious, in fact, that a complex system of overlapping rights to and restrictions on its use had evolved. People guarded these rights jealously, and few indeed were free to use the land as they wished without encumbrance of some sort. The different colonial powers tried to transplant selected aspects of feudalism to the New World. The Spanish, the French, the Dutch, the British each had their own distinctive doctrines about property rights and about how the land should be divided up and parceled out to settlers. These doctrines have left their marks on the landscape, but inevitably they were subverted by the apparently unlimited amounts of land available in the New World.

The early settlers descended on people whose notions of property rights were quite different. Although some tribes grew corn, beans, melons, tobacco, and other crops, often in communal plots, for the most part Indian agriculture was a desultory activity that merely supplemented hunting, fishing, and gathering. Tribes had their traditional hunting grounds, but constant bickering between neighboring tribes indicates that they had a cavalier attitude toward boundaries. Even today the tidy-minded white bureaucrats on the Navajo Indian reservation in northeastern Arizona, who want to draw neat property lines on their maps, are frustrated by the ease and rapidity with which the boundaries of individual grazing areas can change.

The early white settlers in North America, steeped in the European tradition of property rights, simply could not understand the Indian attitude toward land—and vice versa. When a gaggle of strangers, dressed in outlandish garb and speaking in a funny tongue, approached an Indian and offered him a collection of beads, shells, mirrors, knives, and other trinkets in return for his land, the Indian reacted much as you would react if some stranger stopped you on the street and offered to pay you a small sum of money for the privilege of watching the clouds in the sky.

The Spanish and the French, for the most part, did not confuse the Indians with such niceties; they simply took what they wanted and claimed it by right of conquest, but for their feudal overlord, the king, not for themselves. The Dutch and the English were rather more fastidious about paying for what they took (every schoolchild knows the myth about Peter Minuit's purchasing Manhattan Island for twenty-four dollars), but they accomplished almost as little when they treated the perplexed Indian tribes as sovereign states with whom they could negotiate "international treaties." This policy, which the United States continued, has had such unintended consequences as tax-free cigarette sales and gambling casinos on reservations, plus nasty squabbles about hunting and fishing rights.

THE SPANISH

The common American view of the Atlantic seaboard as "colonial America" is fairly myopic because some of the earliest settlements in the United States were along its southern margin, from California to Florida, which was the northern rim of the Spanish empire in the Caribbean. This area was one of the most remote and least valuable outposts of the empire, however, and its limited development was motivated largely by religion and defense: to convert the infidel Indians or to keep hostile groups from occupying areas claimed by Spain.

Florida, for example, was a Spanish possession for nearly two hundred years, but St. Augustine, which was founded more than forty years before the first English settlement at Jamestown, was the only place the Spanish occupied continuously, and modern Florida shows few signs that the Spanish were ever there. The Spanish officially regarded the land as the property of the Indians and did not allow individual Spaniards to acquire title to it, which effectively forestalled settlement.

Fears of foreign aggression inspired many of the territorial activities of the Spanish and, later, Mexican governments farther west. They granted large tracts of land along the frontier, both to get the country settled and to protect it against encroachment: by the French and Yankees in Texas, by Yankees and raiding Indians in New Mexico and southern Colorado, and by the English, Dutch, French, and Russians in California.

The Spanish made four kinds of grants: (1) common grants to Indian pueblos, which merely confirmed the rights of the Indians to land they were already using; (2) individual grants to influential persons, such as government officials, army officers, and men of wealth; (3) community grants to groups of settlers; and (4) development grants to impresarios, who were required by the terms of the grants to bring in settlers. The last three types were controlled, in the early Spanish days, at least, by rigid and comprehensive laws stipulating that the grants must be to unoccupied areas that could be used without injury to the Indians, and with their voluntary consent.

In Texas the Mexican government continued the Spanish practice of granting land in large blocks; an impresario received one square league of approximately 4,428 acres, or roughly seven square miles, for each family he brought in to settle it. The grant made to Stephen F. Austin was larger than the states of Connecticut and Rhode Island combined, and included the better part of more than a dozen contemporary Texas counties. After independence the Republic of Texas granted settlers a stipulated acreage of land that they could stake out on any part of the unappropriated domain, without reference to any system of land survey.

Land division was regulated more closely in the Spanish-Mexican settlements of the upper Rio Grande valley north of Santa Fe, where colonization was based on farming villages, or pueblos, in irrigable valleys. Each pueblo had a grid street pattern, with its church facing a two-block plaza that was used for ceremonies and military drills. The settlers of high rank received house lots facing the plaza, and the house lots of the lower sort were on the back streets. Each settler was given a narrow strip of cropland with access to the irrigation ditch, and had the right to graze livestock on the common pasture near the pueblo.

These features still are clearly visible on aerial photographs of the modern village of San Luis in central southern Colorado (fig. 8.1). South and a bit east of the village is a common meadow of nine hundred acres on which each family has the right to graze ten head of cattle or horses. Farther upstream are the individual strips, 100 to 220 feet wide, that extend ten to fifteen miles back to the watershed separating this valley from the next one. Only that part of the strip at a lower elevation than the irrigation ditch can be cultivated. The higher parts are in dry rangeland of such slight grazing value that no one really knows or cares about the location of the boundaries within it.

Along the coast of California the Spanish established missions and pueblos, built presidios (military posts) to protect them, and granted huge

Fig. 8.1.   An excerpt from an aerial photograph (U.S. Department of Agriculture CWS-5P-33, September 27, 1955) of the village of San Luis in central southern Colorado. All the people live in the village in the upper left corner. The dark area south and east of the village is a commons on which each villager has the right to graze specified animals. The cultivated fields are laid out in long, narrow strips at right angles to the life-giving irrigation ditch.

ranchos for grazing. The missions were dominated by their churches, with living quarters and workshops around the adjacent courtyards (fig. 8.2). Close by were the irrigated fruit orchards, olive groves, vineyards, and cropland where the padres taught their Indian converts how to grow wheat and other crops. The missions were secularized under Mexican rule, but they remain prime tourist attractions in a state with a short history.

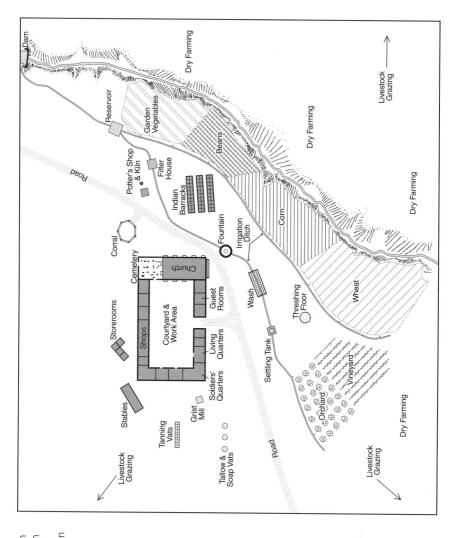

Fig. 8.2. A generalized pattern of land use on a Spanish mission in California. Reproduced by permission of the author from an original sketch by David Hornbeck.

The Spanish and the Mexicans made enormous rancho land grants in California. Under Mexican law the minimal grant was a square league, or about seven square miles, and the legal maximum, which could be exceeded without much difficulty by anyone who knew the right people, was one square league of irrigable land, four square leagues of land suitable for dryland farming, and six square leagues of grazing land, for a total area larger than the entire District of Columbia. The boundaries of these land grants were located wherever the grantee wanted to put them, and they were surveyed, if at all, in the most haphazard fashion, with gaps and overlaps in the marginal areas between ranchos, because such areas were hardly used.

The headaches began when Americans started to pour in to California and demanded clear title to the land before they would settle it. The United States tried to respect the Spanish and Mexican land grants, and it eventually confirmed some of them, but only after extended litigation. The newcomers were unwilling to settle on land that was still under litigation and might later be taken away from them after they had developed it, so the settlement of some areas was long delayed.

Most of the Spanish and Mexican land grants were in just those parts of California that have attracted the largest numbers of people: the areas around San Francisco Bay, Monterey Bay, Santa Barbara, and the Los Angeles Basin; there was also a scattering around Sacramento. These large land grants dating from the colonial era have strongly influenced urban growth in California even in recent years. Until after World War II, for example, the ninety-thousand-acre Irvine Ranch blocked the growth of greater Los Angeles southeastward into Orange County. In a sense the old land grants have been a boon to developers, however, because new technology, which permits residential development of hilly land previously good only for grazing, requires large blocks of land, and it has been easier for developers to acquire such blocks from old land grants that have never been divided than it is to try to piece together blocks of suitable size by acquiring many small tracts.

THE FRENCH

The French government tried to transplant a vestigial form of feudalism, the seigneurial system, to its possessions in Canada. It granted large acreages of land to seigneurs, who were expected to bring in tenants to settle and work the land. In return, the seigneur had certain feudal rights over his tenants: to receive token rent payments from them, to require them to use his mill for grinding their grain, to demand various work services from them. The feudal elements of this system quickly became irrelevant in the New World, where land was available in abundance for tenants who felt oppressed and wanted to move, but the long-lot system of land division associated with seigneuries remains vividly imprinted on the landscape of rural Quebec and of other parts of North America that were settled by French-speaking people.

The French settlers were more interested in the fur trade than they were in farming, and the seigneuries were laid out to give maximum access to the rivers that were the main routes to the interior. Each seigneury had a fairly narrow frontage on the river, but extended far back from it. Each was subdivided into long narrow strips that were only 350 to 600 feet wide but ten times as deep. At the back end of the strips a road ran parallel to the general course of the river, and this road provided the frontage for a second range of strips when the first range was fully occupied.

The long-lot system of land survey was cheap and easy, and it gave each farm equal amounts of each kind of soil on the floodplain, the terraces, and the interfluves. It gave access to a transportation artery (river or road) for a maximum number of farms. Each family could live on its own farm but still be close to neighbors. Each family developed its farm progressively, by clearing the forests near the farmstead first and leaving the more remote parts until later (fig. 8.3). Some farmers built isolated field barns at the far ends of their farms to reduce the labor of hauling crops in to the farmstead and manure back out to the fields.

The long-lot system of land survey has some disadvantages. Few rivers follow straight lines; the twists and turns of a meandering river complicated the survey, the "long lots" inside narrow bends were likely to be bobtailed, and considerable litigation was virtually inevitable if the river changed its course. One of the most serious drawbacks of the long-lot survey system was related to inheritance. Farms were split right down the middle when the farmer died, and farms that were already narrow became far too narrow when they were divided equally among the children, especially when families were as large as they often are in French Canada.

Despite these drawbacks, the French carried the long-lot survey system with them wherever they settled in North America (fig. 8.4). They remained more interested in fur trading than in farming, and except in Louisiana the settlements that developed beside their trading posts were small and grew slowly. The principal mark they have left on the land is the remnants of

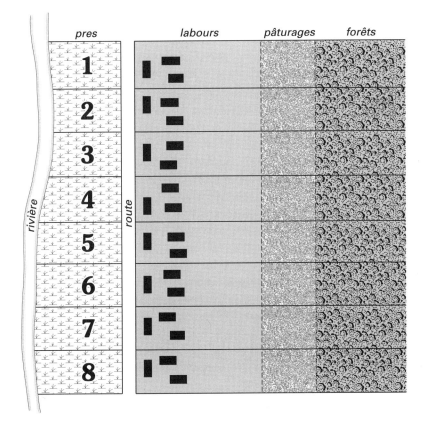

Fig. 8.3. Long-lot farms laid out at right angles to a river in French Canada. Each farm has low-lying meadowland near the stream, a farmhouse and two outbuildings on the road, cultivated land near the farmstead, pasture farther back, and a wooded area at the back of the farm.

their land-division system, which are still visible on modern topographic maps of Detroit, of Green Bay, and of Vincennes, Indiana, as well as on the topographic maps of Louisiana, which are better known.

THE DUTCH

The Dutch also tried to transplant feudal ideas to their early settlements in the Hudson River valley north of New York City. They were even less successful than the French, but they did leave place names such as

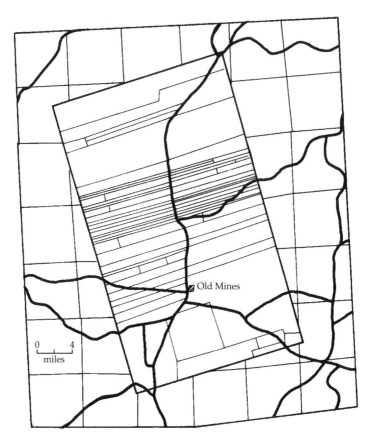

Fig. 8.4.   Land division and principal roads in the area around Old Mines, north of Potosi, Missouri. A large rectangular Spanish land grant was subdivided into French long lots. The land outside the Spanish grant is surveyed by the American township-and-range survey system. Reproduced by permission from Russel L. Gerlach, *Immigrants in the Ozarks: A Study in Ethnic Geography* (Columbia: University of Missouri Press, 1976), 21.

Philipse, Rensselaer, Roosevelt, Schuyler, Stuyvesant, van Cortlandt, Vanderbilt, and Van Wyck. Dutch patroons were granted sixteen miles on one side of a navigable river, or eight miles on both sides, and as far back as they could occupy the land, on condition that they establish a colony of at least fifty adults within four years. They could select any unclaimed stretch of the river, and they could move to another stretch if they changed their minds about the first one. Their tenants had to pay one-tenth of their farm products plus a cash rent, and also had to render services such as woodcutting, hauling, and the repair of roads and buildings.

"In a new and promising country where the natives were friendly, the transportation easy, the land fertile, and all other conditions favorable," said Maud Wilder Goodwin in *Dutch and English on the Hudson*, this attempt to impose the feudal system produced "a sparse and sullen tenantry, an obsequious and careworn agent, an impatient company, and a bewildered government." The patroonship of Rensselaerswyck, across the Hudson River from Albany, persisted for more than two centuries, but no other patroonship lasted for more than a generation or two. In 1638 the Dutch abandoned patroonship in favor of smaller grants to individuals, but few of their compatriots saw fit to forsake the comforts of Holland to accept them.

An English fleet commanded by the duke of York captured New Amsterdam from the Dutch in 1664 and renamed it New York, but the new English rulers continued the Dutch tradition of granting land in huge manors, many of which dwarfed the plantations of the South. It was assumed that these manors would support a landed gentry who would rent out the land in smaller parcels, but most newcomers preferred to settle on free land in the other colonies. Eventually the manors were divided into small farms, but they strongly influenced the spread of settlement in the United States because the large blocks of privately held and sparsely populated land along the lower Hudson River discouraged movement along the easy water-level route to the Great Lakes and the interior of the continent until Yankees moving westward from New England pulled an end run and swept around them to the north.

THE ENGLISH

New England

Apart from the Dutch settlements in the lower Hudson River valley and some efforts by Swedes on the Delaware, the English set the policies for land division along most of the Atlantic seaboard of the United States. The very name *New England* is appropriate because most of the settlers actually did come from England, at least in the early years, whereas areas south of the Hudson were settled by a more diversified group of people drawn from many parts of northern and western Europe.

The early New Englanders were a fairly stiff and intolerant lot who brooked no nonsense, as is evident in their system of land division and settlement. Before 1629 a few tracts of land in New England were granted to individuals, but for almost a century thereafter land was granted only to groups of worthy men, or proprietors. The basic unit was the township or town, a more or less compact block of land whose ideal size was six miles square (fig. 8.5). New grants were contiguous to towns that had already been settled, to ensure that settlement advanced inland in serried tiers, but gores of no man's land were sometimes left between old and new towns when grants were made.

Near the center of their grant the proprietors laid out a village with a meetinghouse (which doubled in brass as church and town hall) and individual house lots facing on a town common or village green. The fields and meadows around the village were divided into strips and parcels of varying size. Each farm consisted of a scatter of these strips and parcels, their number and size depending on the size of the family and the wealth and property the family had brought with it, or some other measure of its value to the community. At first not all the land was needed, and it was held in common for future allocation to the offspring of the original settlers or to new settlers when they joined the community.

When a town was fully settled and more land was needed, a group of its members would hive off and obtain a new block of land at the margin of the settled area. The repetition of this process produced the New England pattern of nucleated villages and fragmented farms. Satellite settlements in the more remote corners of the larger towns often adopted a modified form of the town name; the town of Eden, for example, might have places with such names as Eden Center, Eden Corners, and Eden Junction, or, less imaginatively, North, South, East, or West Eden.

The development of nucleated villages and the division of the land into compact towns were essential to the social, educational, and religious well-being of the stern New Englander, who was more concerned with preserving a way of life than with amassing this world's goods; many towns, in fact, forbade the sale of land to outsiders without approval at the town meeting. The sanctimonious New Englanders were firm believers in original sin; they knew that they themselves were wicked, and they assumed that everyone else must be just as sinful as they were. They wanted to keep everyone in the village because it was more difficult for people to misbehave when they were living side by side under constant surveillance. Those who lived on isolated farms were much more easily tempted to develop disrespect for authority, to indulge in strong language and stronger drink, and even to go so far as to accept bad weather and poor roads as excuses for failing to attend church on the Sabbath.

The New England system began to break down under pressure from

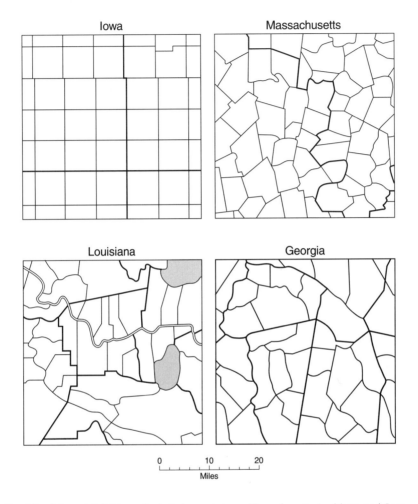

Fig. 8.5. Minor civil divisions of selected areas in Iowa, Massachusetts, Louisiana, and Georgia that show the effects of different land-survey systems. Many civil townships in Iowa are township-and-range-survey townships. The towns of Massachusetts are close in size and have straight boundaries but irregular shapes. The wards of Louisiana are more or less at right angles to the principal waterways. The militia districts of Georgia reflect a metes-and-bounds survey system.

people with less interest in God and more in Mammon. Acrimonious quarrels over the division of land broke out in many towns, and groups of dissidents obtained their own grants and formed new towns. Toward the end of the seventeenth century, large blocks of land on the frontier were sold to people of wealth or prominence who never intended to settle the land themselves, but planned to subdivide it and sell it at a profit to individual settlers. In time even the states started to look on the sale of their western lands as a source of revenue.

Pennsylvania

King Charles II, after his restoration to the English throne in 1660, paid off some of his political debts by granting huge tracts of land in the New World to proprietors, who were authorized to dispose of these vast private estates as they saw fit. Different proprietors naturally had different notions of how they should dispose of their land. The proprietors in Pennsylvania, Delaware, and North Carolina sold their land in small blocks, but in Maryland, Virginia, and South Carolina the proprietors granted large estates, manors, or plantations. In New Jersey the true and absolute lords proprietors of the entire province, Sir George Carteret and Lord John Berkeley (or their heirs), sold their grant intact; within two decades it had been divided into two parts, had more than two dozen proprietors, and had received a legacy of confused land titles that required almost a century of litigation to unravel.

The most successful proprietor was the gentle genius William Penn, who in 1681 was given absolute title to all land between the fortieth and the forty-third parallels, and five degrees of longitude west of the Delaware River. Penn's dispute with his fellow proprietor Lord Baltimore of Maryland over the location of the southern boundary was not settled until 1767, when Charles Mason and Jeremiah Dixon were brought from England to survey the line at 39°43′26″N that is the traditional boundary between the North and the South, the Mason-and-Dixon line. Penn advertised widely and sold land to all comers. He planned to lay out the land in neat squares to encourage community settlement, but this idea failed to appeal to individual buyers, and much of the colony was settled without any regular system of land survey.

Penn's colony became one of the most important culture hearths in the United States. Southeastern Pennsylvania contains the only large area of really good farmland east of the Appalachian uplands, and it supported a prosperous agricultural economy, thriving small towns, and the bustling seaport of Philadelphia. Before the Revolutionary War this city was the principal port of immigration to the American colonies. Many stolid, industrious Germans, later to become known as the Pennsylvania Dutch, settled west of Philadelphia and applied their expertise to the fertile soil or in small factories. The restless, cantankerous Ulster Scots pressed on westward to the frontier, but in passing through the German settlements they learned many skills (such as farming the land and building log cabins), and acquired tools and implements (such as the Kentucky long rifle and the Conestoga wagon) that they used in taming the western wilderness.

One of William Penn's most enduring monuments is the widely copied plan for his City of Brotherly Love, which had a grid pattern of streets. The east-west streets of Philadelphia, which are perpendicular to the Delaware River, were named after trees and flowers. They intersect at right angles the north-south streets paralleling the river, which are numbered consecutively away from it. The two principal streets of Philadelphia, Broad and High (now Market), meet at a central square formed of rectangular corners cut out of the four adjoining blocks. The city hall occupied this square in Philadelphia, but the county courthouse graced it in other early cities of southeastern Pennsylvania. It is hard for a contemporary American, who takes grid street patterns for granted, to realize how attractive this layout must have seemed to immigrants who knew only the "cowpath" street patterns of cities in Europe. The "Philadelphia plan," including the central square and system of street names, has been widely copied in the eastern United States, and most Americans take it for granted as the only proper way to lay out a city.

Metes and Bounds

The systems of land division and settlement in Pennsylvania and the colonies to the south were far more discriminating than those in New England, but the sobriquet "indiscriminate location" was hung on them, perhaps by envious New Englanders, and it has clung. In New England, land was granted in solid blocks next to areas that had already been settled, which ensured that settlement advanced inland in compact tiers, but forced settlers to take the poorest land along with the best. Many groups got stuck with some sorry pieces of real estate.

Settlers in the southern colonies labored under no such constraints. Each person could select the best land that no one else had claimed, and the poorer land was bypassed. The result was an extremely irregular and unsystematic pattern of land division that could produce conflicting claims and protracted litigation over title to the land unless the land was surveyed with a degree of skill that few colonial surveyors possessed (fig. 8.6).

In the southern colonies a prospective settler obtained a land warrant

Fig. 8.6. Land-survey patterns in the Natchez district of Mississippi, including one area of conflicting claims. Reproduced by permission from Sam B. Hilliard, "An Introduction to Land Survey Systems in the Southeast," in *Geographic Perspectives on Southern Development*, edited by John C. Upchurch and David C. Weaver, Studies in the Social Sciences, vol. 12 (Carrollton: West Georgia College, 1973), 13.

from the authorities. This warrant stipulated the number of acres the settler was entitled to claim and settle, but it did not specify their location. Settlers selected likely looking tracts of unappropriated land, preferably far enough from other settlers to provide reasonable assurance against possible claims of encroachment. They laid out their claims, had them surveyed, and presented a copy of the survey, together with the warrant, to the proper official. After a specified period of time, if no valid protest were raised, they were issued a patent or deed to the land (fig. 8.7). This patent was no better than the survey on which it was based, and colonial surveyors, with all due respect to George Washington, were notorious for their bungling and incompetence.

Surveying still is far from an exact science. As late as May 28, 1988,

Michael Winerip reported in the *New York Times* that an entire subdivision of Hartford, Connecticut, had been surveyed incorrectly when it was laid out in 1976, and its good citizens were horrified to learn that their fences, their shrubs, their trees, their tulip beds were on land that belonged to their neighbors.

"It's the worst thing that could have happened except for a flood or a tornado," said the town councilor, who lived in the area.

"Sue the surveyor!" screamed the people.

"Unfortunately," said the town engineer, "he's dead."

Early surveyors identified property boundaries by stating compass directions and distances to and from trees, stones, creeks, spikes, stakes, and other transitory features. A recent description of the boundaries of a single property in Brown County, Ohio, reads, in part:

> beginning at a stone at the original northeast corner . . . thence south 6 deg. 20 min. west a distance of 774.32 feet to an iron rod . . . thence south 11 deg. 42 min. west a distance of 406.94 feet to an "X" marked on a rock in the centerline of White Oak Run . . . thence south along the centerline of said creek with the meanders thereof to a spike in the center of the bridge . . . thence along the remnants of an old fence north 16 deg. 24 min. west a distance of 100.45 feet to a sugar tree . . . a large beech . . . a wooden stake . . . a stone near an elm . . . a large buckeye near a wine cellar.

In fifty years it might be tough to repeat this survey.

The careless administration of grants and titles to inaccurately surveyed properties with poorly marked boundaries proved a wonderful source of income for lawyers; for example, Elias Boudinot of New Jersey complained to the first Congress on December 27, 1790, that more money had been spent in lawsuits over land claims in New Jersey than the entire state was worth. And some boundaries still have not been carefully surveyed; even today some county lines on topographic maps of parts of Kentucky carry the notation "indefinite boundary." It is estimated that the owners of a considerable acreage of land in that state pay no taxes whatsoever because they tell the officials of each county that their land is on the other side of the county line.

With an empty wilderness stretching apparently without end into the interior, the colonies and later the states were only too happy to pick up a bit of change by selling warrants for land, or to use warrants instead of cash

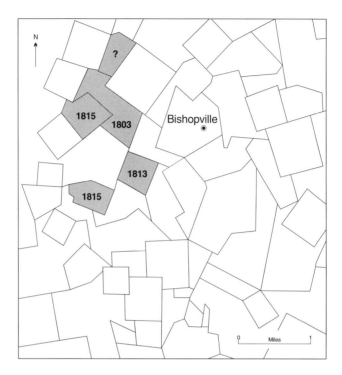

Fig. 8.7. Original land grants under patents from the state of South Carolina in the area around Bishopville, South Carolina. Most of these grants were made between 1770 and 1789. The shaded areas were acquired by John Baxter Fraser in the year shown on each parcel. Based on a map compiled by J. F. Stuckey and inscribed by Joseph H. Parkman, with permission of Margaret DesChamps Moore.

to pay off their financial and military obligations. Speculators were not above selling the same piece of land to several different purchasers, and some officials were little better; twenty-four Georgia counties whose total area was only 8,717,960 acres made land grants totaling 29,097,866 acres. Furthermore, a person with no legal title to land could squat on it and claim preemption rights on the basis of occupancy and development; in 1779 the Kentucky land court allowed anyone who had settled on the land to claim four hundred acres, with an additional thousand acres if the settler had improved it.

THE AMERICANS

The Old Northwest

Problems of conflicting land claims were so serious in the seaboard colonies, and were becoming so threatening west of the Appalachians, that the founding fathers determined they must be solved before settlement could be allowed "on the western waters." Their largest block of unclaimed land was the Northwest Territory, the area north of the Ohio River, east of the Mississippi, and south of the Great Lakes, which the Virginia militia had captured from the British during the Revolutionary War. This Old Northwest became the nucleus of the original public domain of the United States when the Commonwealth of Virginia ceded it to Congress in 1784; other states relinquished some rather extravagant claims to this territory, and to western lands farther south, which were also made part of the original public domain.

Congress appointed a committee, under the chairmanship of Thomas Jefferson, to prepare an overall plan for the systematic survey and subdivision of the public domain before it was alienated to intending settlers. The public domain consists of all land directly owned by the government; alienation is the process by which the government transfers the ownership of public land to private citizens. It is always understood that the government reserves the right of eminent domain, the right to take back private property if it is needed for public use.

The Jefferson committee's proposal, as modified and adopted by Congress in the Ordinance of 1785, is variously known as the Congressional, the General Land Office, the National Land, the Public Land, or the Township and Range Survey System; these names are used interchangeably. The fundamental features of the system, whose details are belabored in many introductory geography textbooks, are: (1) the land had to be surveyed before it could be alienated and settled; (2) the survey lines were oriented in cardinal compass directions; and (3) the land was divided into townships six miles square, which were subdivided into sections one mile square, or 640 acres (see fig. 8.5).

The young American nation used the state of Ohio as a laboratory for experimenting with different kinds of rectangular survey systems. In 1785 Chief Geographer Thomas Hutchins ran the first survey line from Pennsylvania forty-two miles west into Ohio before his party encountered a group of unfriendly Indians who offered them free haircuts, and they decided to beat a hasty retreat to Pittsburgh.

Unfriendly Indians were merely one of the routine hazards of survey parties. They had to hack their way through thick forests to run their lines across rugged country, and were regularly subject to attack by wild animals, snakes, mosquitoes, and malaria, on top of all the usual vicissitudes of wind and weather to which anyone is exposed who works outdoors. And they had little time for pinpoint accuracy or niceties of detail, because settlers clamoring for land were hot on their heels.

Ohio has ten distinct survey tracts, which differ in the size of their townships, in the orientation of their survey lines, in the numbering of their townships, ranges, and sections, and in the way they were alienated (fig. 8.8). Their diversity is reflected by the boundaries of minor civil divisions in the west central part of the state.

Connecticut reserved a large tract, known as the Western Reserve, in northeastern Ohio when it ceded its claims on other western lands to the federal government. The area known as the First Seven Ranges, west of the West Virginia panhandle, was the first area surveyed under the new system, and it became obvious immediately that some modifications were necessary. The Ohio Company of New England bought a million and a half acres in the hilly southeastern corner of Ohio directly from the federal government in 1787, and a year later a group from New Jersey headed by John Cleve Symmes bought the tract between the Miami and Little Miami Rivers in the southwestern part of the state.

The Commonwealth of Virginia retained the land between the Little Miami and Scioto Rivers as the Virginia Military District, where Revolutionary War veterans could redeem the land warrants they had received for their military service. Unlike the rest of Ohio, the Virginia Military District does not have a rectangular survey system, and even though it has only one-sixth of the state's area it has enjoyed more litigation over property boundaries than the rest of the state combined.

The Greenville Treaty Line, which cuts across Ohio at an angle of ten degrees north of east (see fig. 8.8), dates from 1795. After the battle of Fallen Timbers, General Anthony Wayne assembled the leaders of the defeated Indians at Greenville and ordered them to sign a treaty in which they relinquished all claims to land south of this line. In return, they were given permanent title to the northwestern corner of the state. In this instance, "permanent" lasted for twenty-three years, which was quite a long time for Americans to honor a solemn treaty they had signed with a group of Indians.

By 1818, when the Indians were expelled, the bugs had been worked

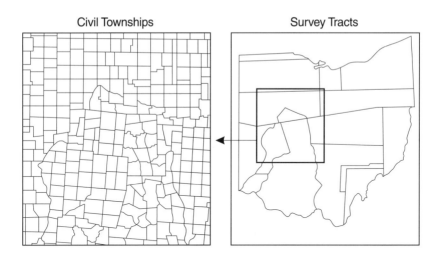

Fig. 8.8. Civil townships and survey tracts in the state of Ohio. The township lines in the northern part of the inset area, which follow the cardinal compass directions of the congressional land-survey system, contrast with the township lines of the Virginia Military District in the south central part.

out of the system in other parts of the state, and the rectangular survey system that was used in northwestern Ohio became the standard that was subsequently extended to most of the United States west of the Appalachian uplands. In many places the straight north-south and east-west lines of the survey are followed by the boundaries of states, of counties, and of minor civil divisions within counties (see fig. 8.5). The lines of the survey system determine the boundaries of most properties, and all but the most important highways follow property boundaries instead of cutting across them. The rural Midwest is crisscrossed by straight "section-line" roads spaced at one-mile intervals that intersect at right angles, and the survey is why many American farmers reckon their land in terms of sections of 640 acres, quarters of 160 acres, and "forties," or quarters of quarters.

The rectangular survey system did not, as is often but erroneously assumed, produce a neat and tidy checkerboard of square 160-acre farms divided into four square 40-acre fields. Careful examination of almost any area in the Midwest reveals a far more complex mosaic of farms and fields. In twelve square miles in north central Indiana, for example, I found only nine square 40-acre fields where the checkerboard theory predicted there would be 192, and Norman J. W. Thrower found an almost identical dis-

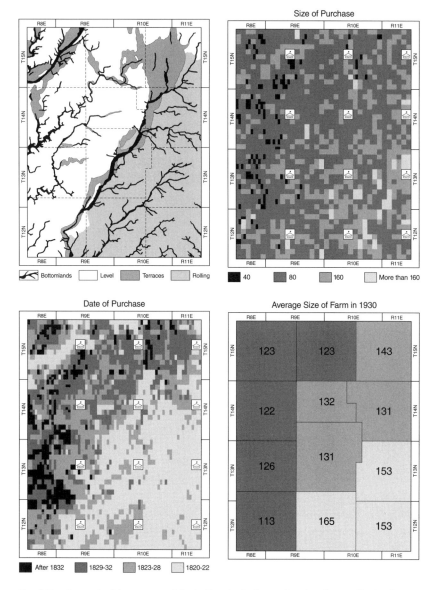

Fig. 8.9. A panel of four maps of Rush County, Indiana, showing the continuing importance of original land surveys and the date and size of original alienation parcels. The map of the size of parcels originally purchased shows that the original purchase units were smaller toward the northwest. The map of the date of original purchase shows that the smaller parcels were purchased later. The map of land quality shows that rolling land was purchased before level land, which had drainage problems. The map of farm size in 1930 shows that farms in the northwestern part of the county remained smaller than farms in the southeastern part a century after the land was originally alienated. Compiled from maps in Wayne E. Kiefer, *Rush County, Indiana: A Study in Rural Settlement Geography* (Bloomington: Indiana University Department of Geography, 1969), and used with the permission of the author.

crepancy in Ohio. The section-line roads do outline squares, but within the square sections the layout of fields is strongly influenced by the needs of the farming system, and most fields are rectangular rather than square.

Pros and Cons

Norman Thrower, in his monograph *Original Survey and Land Subdivision*, has described the effects of different survey systems on transportation lines and on administrative, property, and field boundaries. He used two sample areas in Ohio that were similar in their landforms, lithology, soils, climate, crops, and population density. Area S was surveyed systematically under the township-and-range survey system, and area U, in the Virginia Military District, was laid out unsystematically, with no overall plan.

The roads in area S cut straight across the countryside along section lines in a neat rectangular grid, but in area U the spiderweb road network followed the lie of the land without regard to survey lines. Area S had a denser road net than area U and many more large and expensive highway bridges, which is another way of saying that larger parts of area U were farther from public roads, and farm driveways had to be longer. The size and shape of the irregular fields in area U varied far more than those of the rectangular fields of area S.

Many original survey lines were used as administrative (county, township, school district) boundaries and property lines in area S, but few were so used in area U. The property units in area U that were in two or more political jurisdictions were a real headache for taxation purposes (both collecting and paying). Furthermore, area U had suffered far more litigation over property boundaries than area S, and the property-boundary lines in area S had changed less than those in area U.

Property-boundary lines show astonishing tenacity and permanence. Thrower found a total of 649 miles of property boundaries in 1875 and 1955 combined, of which 463 miles, or 70 percent of the total length, were the same in both years. In *Rush County, Indiana*, Wayne E. Kiefer said that many modern property lines are the boundaries of the units that were purchased from the General Land Office when the land was originally alienated in the 1820s and 1830s, and the size of contemporary farms is closely related to the size of the parcels of land that the first settlers purchased (fig. 8.9).

In recent years it has been high fashion to criticize the township-and-range survey system, whose strengths and weaknesses have been identified

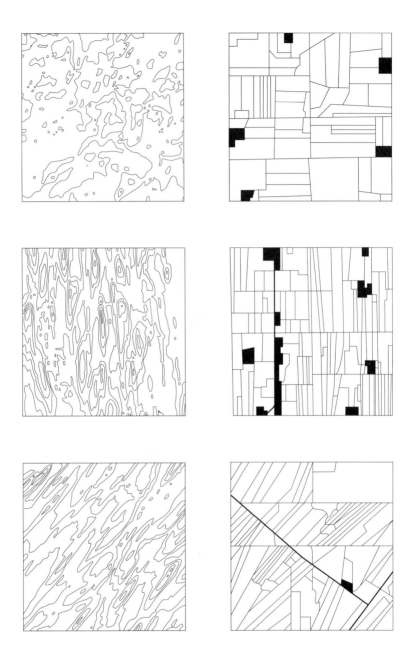

Fig. 8.10. Contour lines (left) and field orientation (right) in three parts of a drumlin area in Wisconsin. The fields in the ground-moraine area (top pair) have no dominant orientation. In the two bottom pairs, the fields have the same orientation as the drumlins. From Charles W. Collins, "The Influence of Drumlin Topography on Field Patterns in Dodge County, Wisconsin," *Wisconsin Academy of Science, Arts, and Letters* 59 (1971): 55-66, and used with the permission of that academy.

by Thrower. Certainly that system has necessitated greater expenditures for such rural services as highways, telephone and power lines, mail delivery, and school-bus routes. In rough, dissected areas it has posed problems in the layout of transportation lines, and it has encouraged farmers to lay out their fields against the grain of the country, to the dismay of conservationists (fig. 8.10).

Nevertheless, the township-and-range survey system may be the best system of land division ever invented. It provides an excellent frame of reference for orientation, and it conveys a sense of neatness, order, and stability. It has obviated an enormous amount of litigation by facilitating a brief but precise description of the exact location of any tract of land, and it permitted the alienation of the public domain to individual owners in compact units of similar size.

The Alienation of Public Lands

In 1785 the new United States was a pitifully poor nation, and its western lands were one of its principal assets. The early land policies of the young republic were designed to obtain revenue from the sale of these lands rather than to encourage their settlement, but for decades the halls of Congress echoed with debates about the minimal price at which land should be sold and the minimal acreage that a buyer should be required to purchase.

Some congressmen from the East managed to find virtue in all the blunders that had been made in settling New England. Revenue could be raised most efficiently, they argued, by selling land only in large blocks to men of substance, who could bear the costs of subdividing, reselling, and settling it. Dividing the land into small parcels would appreciably increase the cost of surveying it, and parcels would fetch a lower price if they were so small that wealthy buyers could not be bothered to bid on them.

Furthermore, they said, it would be more difficult for prospective purchasers to select only the better land and bypass the poorer if the land were sold in large blocks, and the danger of Indian attack would be reduced if settlements were grouped on blocks of land instead of being scattered through the wilderness. Underlying all these rationalizations, although it was rarely articulated, was the fear that factories in the East would lose their supply of cheap labor if workers were lured westward by the availability of small blocks of land at low prices.

The congressmen from the West would have none of such nonsense.

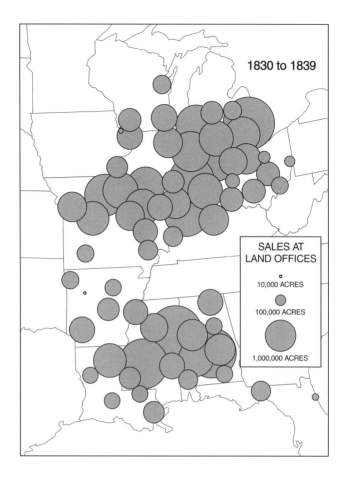

Fig. 8.11. Land sales at land offices in the United States between 1830 and 1839. Reproduced by permission from H. J. Walker and W. G. Haag, eds., *Man and Cultural Heritage*, Geoscience and Man, vol. 5 (Baton Rouge: Louisiana State University School of Geoscience, 1974), 78.

They argued that settlers were performing a patriotic service when they tamed the wilderness and advanced the frontier. Land should be free, they said, or at least it should be available in tracts so small and at prices so low that every person who wanted a farm could afford the "threshold price" (the minimal acreage of land multiplied by the minimal price). They gradually managed to hammer down the minimal sales unit from 640 acres in 1785 to 320 acres in 1800, 160 acres in 1804, 80 acres in 1820, and 40 acres from 1832 until 1862, when the Homestead Act gave 160 acres free to anyone who would live on the land and cultivate it for five years.

Before 1820, federal government land sales had to compete with those of the states, which had land of their own to sell, and with holders of military warrants, most of whom wanted only to convert their warrants into cash as quickly as possible. When these sources of cheaper land began to dry up, however, the minimal purchase price for federal land was reduced from $2.00 an acre, where it had been set in 1796, to $1.25 an acre. In other words, the threshold price for a farm was reduced from $1,280.00 (640 acres at $2.00 an acre) in 1796 to $100.00 (80 acres at $1.25 an acre) in 1820 and to $50.00 (40 acres at $1.25 an acre) in 1832, where it remained until 1862, when the secession of the South finally allowed Congress to pass the Homestead Act. For comparison, the wage for a farmhand in Ohio ranged from $5.00 to $15.00 a month and board, so after 1832 the minimal cost of a farm worked out at somewhere between three and ten months' wages for a farmhand.

The Homestead Act was largely irrelevant in the Midwest, where most of the land had been bought and paid for before the act was passed in 1862. Much of the eastern Midwest was purchased during the land boom of the 1830s (fig. 8.11), and most of Iowa and Wisconsin during the land boom of the 1850s (fig. 8.12), when the minimal purchase unit was only 40 acres. By 1862 most of the land that was still available for homesteaders was in Minnesota, the Dakotas, Nebraska, and Kansas.

In the early days few people bought more than the minimal acreage required by law, because farm machinery was primitive, and a man with horses could not cultivate more than 40 to 80 acres even if he had a large family of husky sons. Some settlers bought larger blocks, but most of the eastern Corn Belt was alienated in parcels of 40 to 80 acres, and even today few farmers own as much as 160 acres, although they farm far more because they rent land from their neighbors.

The areas that once had the smallest farms still have the smallest farms, and the areas that began with the largest farms still have the largest farms. Farms in Ohio and Indiana still are smaller than farms in areas farther west that were opened up later with the kinds of improved machines that had been developed on Ohio and Indiana farms. The largest farms in Ohio are in the Virginia Military District, where people used military warrants to acquire land and did not have to pay cash for it.

The geographic stability of property boundaries and farm size suggests

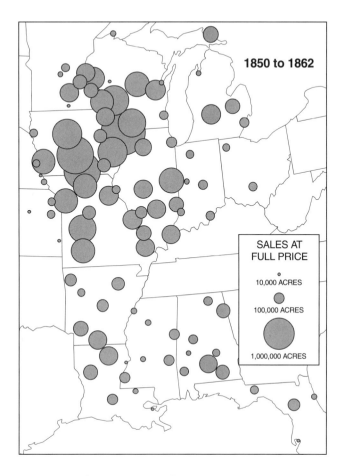

Fig. 8.12. Land sales at full price at land offices in the United States between 1850 and 1862; sales at graduated prices are not shown. Reproduced by permission from H. J. Walker and W. G. Haag, eds., *Man and Cultural Heritage*, Geoscience and Man, vol. 5 (Baton Rouge: Louisiana State University School of Geoscience, 1974), 79.

that the size of contemporary farms in the Midwest may be a function of the acreage that the first settlers originally obtained from the federal government more than a century ago—yet another manifestation of the fundamental geographical principle of *prior potior* (the earlier has the greater influence).

In addition to the Homestead Act, Congress passed a host of other land laws. Some granted land for worthy purposes such as planting trees in semiarid areas or draining swamps. The Graduation Act of 1854 authorized the sale of land at reduced prices if it had been on the market for ten years or more. The Morrill Act of 1862 gave land to each state to establish a land-grant agricultural college; the settled states in the East were permitted to claim land in the West. The spread of settlement was facilitated by land grants for internal improvements such as the construction of canals and railroads. Improved transportation increased the value of the land, which the grantees sold to pay the costs of construction, and the grantees encouraged settlement along their lines to ensure a steady volume of traffic and revenue.

The careful records of land survey and alienation that were kept by the United States General Land Office have been a veritable gold mine of information for geographers and other scholars. Wayne Kiefer and C. Barron McIntosh compiled maps showing the spread of settlement by mapping the date at which each individual parcel of land was alienated (fig. 8.13; see also fig. 8.9). The compilation process is extremely slow and tedious, but it can produce spectacular maps that give remarkable insight into the way the settlers perceived the character of the natural environment they occupied. Geographers have also used surveyors' notes, which record the vegetation along their survey lines, to reconstruct the vegetation that existed at the time of the original survey.

THE CANADIANS

Before 1763 most of the people of Canada lived in the St. Lawrence valley between Montreal and Quebec, where the land was divided into long lots. The French settlers were more interested in the fur trade than in farming, and the long-lot system of land division enabled everyone to live near the rivers, which were the principal routes to the west. The French authorities established forts at key nodes in the trading routes, but they permitted settlement only in the immediate vicinity of these forts. Small areas were laid out in long lots along the Detroit River at Windsor, in the Red River valley near Winnipeg, and in other places scattered across western Canada, but these areas are little more than curiosities.

The British did little to encourage settlement after they seized control of Canada from the French in 1763, but the population of southern Ontario increased rapidly after the Revolutionary War, when Loyalist refugees (Tories) from the American colonies began flocking to the safety of the British forts at Niagara and at Detroit. They were joined by a small number

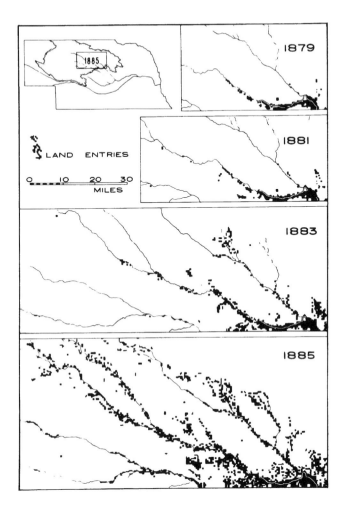

Fig. 8.13. A series of maps showing land entries in specific years shows how settlement spread in the eastern part of the Nebraska Sand Hills. The concentration of entries along stream valleys is especially striking. Reproduced by permission from C. Barron McIntosh, "Patterns from Land Alienation Maps," *Annals of the Association of American Geographers* 66 (1976): 574.

of Iroquois Indians from upstate New York, and after the war by demobilized British and German veterans who had been given land warrants as rewards for their military service. The demand for land was so great that the land had to be surveyed quickly, and the resulting experimentation produced a remarkable crazy quilt of five different major township systems with no fewer than 166 local variations (fig. 8.14).

The first basic type of survey in southern Ontario was the single-front system. In theory, at least, a single-front township was six miles square, the standard unit of land with which the Loyalists had been familiar in New England. The surveyor laid out a base line six miles long at the edge of a township and drew lines at right angles to the base line to divide the land into twenty-five lots that were one mile deep. Each one-by-six-mile strip of twenty-five lots was called a concession. The surveyor laid out a road allowance sixty-six feet wide for a concession road along the base line.

Then the surveyor laid out five more concession lines parallel to the first and one mile wide, each with its own road allowance. Each concession was divided into twenty-five lots, so the completed township consisted of six concessions, or rows of twenty-five mile-deep lots. Side roads every few miles connected the concession roads. The individual lots were broader and not so deep as the long lots of Quebec, but their shape was an advantage because the settlers were required to build and maintain the roads in front of their lots. Long lots yielded a greater labor supply and were less onerous on individual settlers.

The road allowances were a major improvement over land-survey sys-

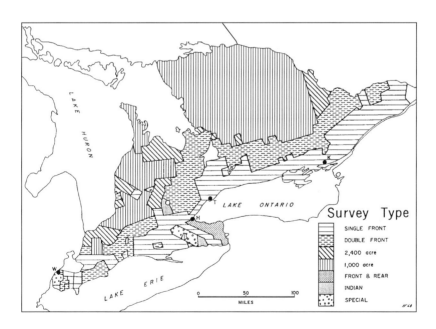

Fig. 8.14. The complex variety of survey systems in southern Ontario. Reproduced by permission from R. Louis Gentilcore, "Lines on the Land: Crown Surveys and Settlement in Upper Canada," *Ontario History* 61 (June 1969): 60.

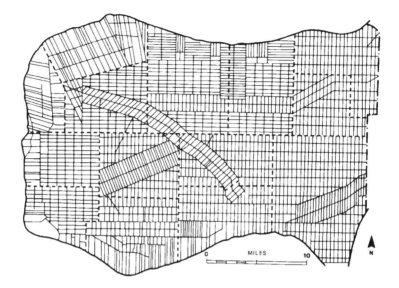

Fig. 8.15. Minor civil divisions in far southwestern Ontario, showing a variety of different doctrines of land survey. Reproduced by permission from J. Clarke and D. L. Brown, "Pricing Decisions for Ontario Land: The Farm Community and the Speculator in Essex County during the First Half of the Nineteenth Century," *Canadian Geographer* 31 (1997): 173.

tems in the United States, in which the construction of roads took land from the individual properties that bordered them. The first township surveys in Ontario included standard town sites with one-acre lots, but not a single town was actually developed according to plan, and this requirement was dropped in 1792.

If in theory the single-front township was six miles square, in practice townships varied in size (six to twelve miles on a side), in number of concessions (four to sixteen), in size of lots (120 to 200 acres), in length of lots (five to two times their width), and in the spacing of their side roads (two to five miles). In double-front townships, lots were laid out on both sides of a concession line, so only every other concession road had to be built. The 2,400-acre and 1,000-acre surveys resembled the American township-and-range survey townships, but a 2,400-acre "section" contained twelve 200-acre lots, and a 1,000-acre "section" contained ten 100-acre lots.

New townships had to respect the boundaries of those that already existed, which necessitated some triangles, parallelograms, and other weird angles, and the complexity was compounded when concessions of narrow farms were laid out at right angles to some of the major colonization roads that sliced diagonally across the countryside (fig. 8.15).

A system that looks chaotic on paper was even worse on the ground because the quality of surveying was inferior. Many survey crews consisted of untrained would-be farmers who received land as payment for their work, and some of them joined the crews in order to pick out the best areas. Many rural roads in southern Ontario have awkward jogs because of surveying errors, and disputes about boundaries were innumerable. Eventually virtually the entire area had to be resurveyed. The new surveys had to respect existing roads and fence lines, no matter how far off they were, because the original survey errors had been built into them.

The rectangular land-survey system of the Prairie Provinces of western Canada has fewer problems, because it is based on cardinal compass directions. Its east-west base lines begin at the forty-ninth parallel, which is the international boundary between Canada and the United States. The parallel base lines are twenty-four miles apart, and they are numbered consecutively from south to north. At six-mile intervals on the base line, meridians are surveyed twelve miles north and south. The convergence of meridians at high latitudes forces these lines to jog every twenty-four miles to keep them six miles apart. The line along which these jogs occur, which is midway between two base lines, is called the correction line.

The base lines and meridians divide the land into nearly square townships. The east-west rows of townships are numbered from south to north, beginning at the international boundary. The north-south rows of townships are called ranges. The ranges are numbered from east to west, starting at one of six principal meridians, so each township has its own township and range number. Each township is divided into thirty-six sections that contain as close to one square mile, or 640 acres, as the convergence of meridians will permit. Each section is numbered, beginning with section 1 in the southeastern corner and zigzagging west and east to section 36 in the northeastern corner. The survey system provides a road allowance every mile east and west, and every second mile north and south.

Part Four

FARM STRUCTURES

Survey lines must be identifiable unambiguously on the ground, and the most obvious manifestations of land-survey systems on the visible land-scape are the fences that mark field boundaries. An observant traveler may sense a survey system, and occasional vantage points may offer a bird's-eye view of the way the land has been divided, but usually you can see survey systems only on maps and aerial photographs. In contrast, you can actually see field boundaries on the ground, and they have piquant tales to tell to anyone who learns to read them aright. Not only do they show the lines that surveyors laid out on the land, but fences, or even the absence of fences, may also send messages about the rural economy and its physical environment.

Fence lines lead to the farmstead that is the focus of farm operations. Barns are the largest and most imposing structures on most farmsteads, and they reflect the nature of the farm economy, but many agricultural activities require special purpose-built structures in addition to barns. Some of these structures are for livestock, some are for crops. Many are at the farmstead, but some are in the fields, and those associated with the collection, processing, and marketing of crops and livestock often are in small towns and villages.

In any given area the variety and diversity of barns and other farm structures can seem overwhelming at first glance, but on closer examination it becomes obvious that they have much in common, because similar groups of people in similar environments have similar needs and make similar decisions that result in similar types of farm that require similar structures. In other words, distinctive regional types of barn and other farm buildings are associated with distinctive regional types of farming, and most farmsteads in most areas have similar suites of structures.

Most areas also seem to have a "normal" size of farm that is appropriate to the local farming system. The tenure and size of farms, whether measured by their acreage or their volume of sales, have important effects on the visible landscape because they influence the range of options within which farmers make decisions. Some literal-minded souls complain that you cannot actually see farm size, but in fact you can, because the size of farms determines the number of farmsteads. An area of 80-acre farms, for example, will have eight farmsteads per square mile, whereas an area of 160-acre farms will have only four. It is silly to go to the effort of trying to estimate the average distance between farmsteads when the average size of farms gives the identical information.

9 Fences and Fields

People put up fences to mark the boundaries of their properties, to sub-divide their properties into manageable units, and to control the movement of their animals—both to keep them out of the cultivated fields where they have no business and to keep them in the grazing areas where they belong. Any one of a great variety of structures can satisfactorily discharge the simple function of enclosing land and livestock. Farmers may settle for barbed wire if they have cattle or horses, but they must use woven wire to confine hogs or sheep. Farmers have used wooden rails, boards painted white, stone walls, hedgerows, strands of electrified wire, even stumps that were uprooted when the land was cleared.

Even ditches may serve as fences. In some reclaimed areas in the Netherlands, for example, cows are kept where they belong by the drainage ditches that control the water level in the soil. Travelers in the polders may be startled by the unusual spectacle of an isolated gate sitting all by itself out in the middle of nowhere until they realize that it merely breaks the line of a water-filled ditch (fig. 9.1).

FENCES IN BRITAIN

Hedgerows

Some property boundaries in Britain may be old indeed, but most of the hedgerows that create what many people consider the typical landscape of lowland England date only from the period between 1750 and 1850, when farmers enclosed the vast open fields to facilitate their use of the new cropping systems and the new breeds of livestock that were associated with the Agricultural Revolution (see fig. 7.8). Fields of five to ten acres were considered the ideal size for the new technology. Farmers dug ditches for drainage, piled the earth in banks beside the ditches, and set live hawthorn

Fig. 9.1. A gate in a Dutch polder where the animals are enclosed by water-filled ditches rather than by fences.

bushes on top of the banks. The archaic English meaning of *quick* was "living, alive," and some people still refer to live hedges as "quickset hedges."

Farmers could rejuvenate their hedgerows by laying or pleaching them, and throughout the nineteenth century hedges remained the cheapest and most effective field boundaries in lowland England. They still dominate the rural landscape, especially in areas of heavy clay soil, and many rural roads are deep green trenches flanked by high hedgerows, offering only fleeting glimpses of the fields on either side. After World War II, the total length of all the hedgerows of England and Wales was greater than one hundred round trips between New York and San Francisco, but roughly one-third of them have been grubbed out since then because they are the relics of a system of land management that has become technologically obsolete.

Hedgerows are expensive to remove, but many farmers consider the cost worthwhile because the hedgerows fossilize fields so small that modern farm machines hardly have room to turn around in them, and farmers must consolidate their holdings into fields of fifty acres or more in order to use these machines efficiently. Farmers also complain that labor has become so expensive that they can no longer afford to lay their hedges by hand, and the available hedging machinery does not do the job properly. Many hedgerows have become derelict and gappy, patched with unsightly boards and wire. They occupy land that could be cultivated, they shade out growth in the adjacent rows of crops, and they compete with crops for plant nutrients. They harbor weed seeds and crop pathogens, and are comfortable denning areas for small herbivorous animals that can devastate crops. They limit the effectiveness of spraying for weeds and insects, and in fruit-growing areas they hinder the flight of pollinating insects. Small wonder that they are disappearing.

The thought of amber waves of grain might make American hearts beat faster, but it makes many English people angry, and they bitterly object to the removal of hedgerows. They sneer about "prairie landscapes," and as one of them said to me, "I don't want the countryside to look like a blooming airstrip." Hedgerows are valuable windbreaks that shelter livestock, they say, and they reduce transpiration and wind erosion on sandy soils. They are ecological reservoirs for birds, insects, bugs, beetles, spiders, wasps, rabbits, mice, hedgehogs, weasels, foxes, stoats, and partridges, and they are corridors that link different wildlife habitats. They harbor predators that prey on pests as well as the pests themselves, and they produce valuable timber, although some foresters say that the value of hedgerow trees for timber production is overestimated. In the final analysis the principal argument of hedgerow advocates is aesthetic; they think hedgerows are more attractive than fences, and they dislike change in what to them is the traditional English landscape of hedgerows. Unfortunately for their cause, aesthetics can rarely compete with economics, and advocates must mobilize all the political muscle at their disposal if they hope to preserve hedgerows.

Dykes

Hedgerows lace together the landscape of much of lowland England, but drystone walls, or dykes, dominate the harsher upland areas of the north and west where sour soils, lower temperatures, and relentless exposure to strong winds and driving rain inhibit the growth of hawthorn and other hedgerow plants (fig. 9.2). They are called drystone dykes because the dykers, who like to boast that they never have to pick up a piece of stone more than once, have fitted the stones together so skillfully that they interlock without the need for any mortar, and some dykes have been standing for centuries. They faithfully reflect the underlying geology, because it costs too much to haul stone any great distance for anything as

Fig. 9.2. Stone walls in a British upland area where the climate inhibits the growth of hedgerow plants.

prosaic as a farm wall, and most of the stone is quarried on the premises, so to speak. The rich, creamy stone of the Cotswolds, the somber gray limestone of Derbyshire, the dark, brooding Millstone Grit of the Yorkshire dales, each adds its own distinctive flavor to the landscape (see fig. 1.3).

In some areas farmers have built head-high banks by stacking alternating layers of stones and sod. The banks quickly become overgrown with grass and shrubs in the mild, moist maritime climate, and they look like nothing more than simple earthen banks, but woe betide the feckless soul who tries to plow through one with a bulldozer.

AMERICAN WOOD AND WIRE

Wood

Which should be fenced, crops or livestock? English common law holds owners responsible if their animals damage someone else's crops, but on the American frontier any unfenced land was considered fair game for livestock, and a body of custom developed under which farmers were required to protect their crops. As late as 1950 a few ranching areas in the West still had unfenced open-range areas, where cattle had the right of way even on state highways (fig. 9.3), and many of the piney woods counties of southeastern Georgia did not have fence laws until 1953. The pressure for these fence laws came not from farmers who were worried about their crops, but from businesspeople who realized that cattle roaming freely on the highways, and the accidents they caused, were bad for the tourist business.

Fencing was simple enough for the early settlers; they merely heaped up the stones, brush, and trees they had cleared from their land into barriers against the animals that roamed freely through the woods around their small cultivated patches. When these rude barricades deteriorated they constructed more-permanent fences of whatever material came most readily to hand, and in the East that usually meant wood.

The first wooden fences were constructed of tree trunks split into ten-foot rails (see fig. 1.4). Farmers interdigitated stacks of six or eight rails at right angles to form a zigzag fence that was so common and so widespread that it acquired a variety of local names, such as worm, snake, split-rail, and Virginia. Sturdy young fellow such as Abraham Lincoln, nicknamed "the Railsplitter" for his prowess, could split up to 150 rails in a day. The recon-

Fig. 9.3.   Unfenced open range in a ranching area in southern Colorado.

structed pioneer village in Lincoln's New Salem State Historic Site, north-west of Springfield, Illinois, has some nice examples of split-rail fences, but nowadays they are rare indeed except in parks and museums or in remote and isolated areas. They have been replaced by more-modern fences when they have burned or rotted away, and they are too expensive and too unstable to be useful for decorative purposes.

It was easy to knock off the top rail of a split-rail fence, and farmers learned to secure their top rails by leaning "rider" rails across the stack at each angle. Some of them bolstered their fences even more securely by driving rails into the ground on each side of each angle, thus creating a stake and rail fence that still was fairly inefficient. The zigzag strip was six to eight feet wide, and the angles harbored weeds and vermin.

The post-and-rail fence was an alternative that used less land and took fewer rails. The farmer cut three or four holes through posts six or seven feet long, set the posts firmly in the ground in a straight line, and inserted rails into the holes. Post-and-rail fences also have pretty much disappeared from the countryside, but they are popular in some suburban areas.

Hedgerows

Settlers from the East were concerned about the lack of wood for fencing when they moved into the treeless prairie grasslands of the Midwest. Some tried ditches and banks of sod, but neither of these worked well, and some experimented with plants they might use for hedgerows. The most popular was the Osage orange, which enjoyed a tremendous boom after 1845. Farmers from Illinois westward to Kansas planted thousands of miles of Osage orange to make thick, thorny hedgerows four feet wide and five feet high that they boasted were "horse-high, pig-tight, and bull-strong." The young plants had to be protected by some other kind of fence until they were three years old, and thereafter they had to be cut back regularly or they would grow into trees fifteen to twenty feet high (fig. 9.4).

The Osage orange boom lasted only a decade; it collapsed abruptly after many plants were killed by the severe winters of 1855–56 and 1856–57. Some prairie areas still have a few long derelict strips of overgrown Osage orange hedgerow that begin nowhere and end nowhere, but most of the old hedgerows have been grubbed out or bulldozed. Hedges in North America have the same drawbacks as hedges in England, and in addition, few hedging plants can successfully withstand the vicissitudes of a continental climate.

Fig. 9.4. Derelict Osage orange hedgerow on the Grand Prairie of east central Illinois.

Nevertheless, the idea of a hedge as a fence that needs little upkeep but provides marvelous shelter for wildlife has beguiled generations of nature-lovers. The Osage orange failed in the Midwest in the 1850s, the honey locust failed in the East in the 1880s, and the multiflora rose, which was widely touted as the solution to all problems in the South after World War II, is hardly mentioned anymore. Perhaps the most successful hedging plants in the United States were the prickly pears that the Spanish padres of the California missions planted around their fields to protect them from the herds of cattle that roamed the surrounding ranchos.

Wire

The effective settlement of the great interior grasslands of North America had to await the invention of barbed wire in 1873 (fig. 9.5). Farmers in the East also used barbed wire to patch up or replace many older rail, board, and stone fences. Three or four strands of barbed wire stapled to wooden posts make a cheap and effective fence for keeping cattle and horses where they belong, but barbed-wire fences will not hold hogs or sheep, which require woven wire (fig. 9.6). Woven-wire fencing, which is

Fig. 9.5. A barbed-wire fence enclosing horses in a ranching area on the Great Plains.

often called hog wire, was developed in the 1850s, but the early version was made of ungalvanized wrought-iron wire that rusted and broke in cold weather. Woven wire did not really become popular until cheap steel became available in the 1890s, but since then most of the fences in the United States have been made of steel wire, either barbed or woven.

In general, the older, established, and more prosperous livestock areas usually are fenced with woven wire, usually topped with a strand or two of barbed wire. Woven wire is more efficient but also more expensive than barbed wire. Barbed wire, cheaper but also more dangerous, is the norm in ranching areas in the West, where the length of fence required per animal is great. Barbed wire also is the usual first fence in areas that are shifting toward a livestock economy, both because it is cheap and because a single strand of barbed wire can conduct an electric current.

A single strand of electrified wire attached to white insulators on flimsy posts is almost invisible, but it is a surprisingly effective fence because animals quickly learn to steer clear of it. Although they are cheap and easy to construct, electric fences are useless during power outages caused by storms or repair work, and they may be grounded by plants or small sticks lodged against them, no matter what the advertisements say. Their current must be

Fig. 9.6. A woven-wire fence enclosing hogs.

adjusted to the amount of moisture in the ground, which is difficult when a fence runs across soggy lowlands and well-drained uplands.

Electric fences perforce are temporary. They may be used to control animals until permanent fences can be put up, but they are most widely used to control rotational grazing, which is the best way to manage a pasture. The farmer divides the pasture into smaller paddocks with electric fences, allows the animals to graze one paddock intensively while the others are recuperating, and then moves the animals to the next paddock. Rotational grazing provides better forage for the animals and also reduces the danger of parasites and disease.

Range managers in the West have developed a novel system for controlling cattle without using fences. They attach an electronic receiver weighing only a few ounces to the ear of each animal; it must be tough enough to survive being chewed on and banged against water troughs and posts. The managers then set up a line of transmitters along the edge of the area in which they wish to contain their cattle. When an animal approaches this line it hears a high-pitched warning sound, followed by a mild electrical shock, and it learns to stay clear. In early trials the shock lasted for a full second, and some animals became so confused that they wheeled com-

Fig. 9.7. A limestone fence post in the post-rock country of north central Kansas.

Fig. 9.8. A rail fence on X-shaped bipods where the solid bedrock is too close to the surface to permit posthole digging.

pletely around and bolted across the line, but shocks of one-eighth of a second seem to work just fine.

Posts

All wire fences must be attached to supporting posts. Steel posts are best because they last the longest, but they also cost the most. The best wood posts are cedar or locust, which will last twenty years or more, whereas pine posts are good for no more than three to five years unless they are treated with chemical preservatives.

Stone posts take the place of wood in the post-rock country of north central Kansas, which once had five million posts of rough, square-edged limestone (fig. 9.7). Wood was scarce and expensive in this treeless prairie area, which is underlain by a remarkably uniform bed of limestone, eight to twelve inches thick, that is soft enough to be worked with hand tools when freshly quarried but hardens when exposed to air. The local farmers learned to wedge off limestone posts eight inches square and five or six feet long that they used instead of wooden posts. One of the advantages of these stone posts was their resistance to the fearsome wildfires that raced across the dry prairie grasslands toward the end of summer.

Steel posts can be driven into the ground, but wooden and stone posts must be set in specially dug postholes, and digging postholes is an onerous task because a manual posthole digger is one of the most awkward devices ever invented. In some areas digging postholes simply is not feasible. In areas of glacial deposition, stones and boulders may be distributed through the soil so erratically that trying to dig a straight line of postholes at regular intervals is an exercise in frustration. In other areas the solid bedrock is so close to the surface that there is no topsoil in which to dig postholes.

Some farmers use bipods where they cannot dig postholes (fig. 9.8). A bipod consists of two posts nailed together in an X shape. The bottoms of the posts rest on the ground instead of being inserted into it. Farmers in Quebec hang wooden rails on loops of wire suspended from each bipod. On the western Great Plains, ranchers attach barbed wire or wooden rails to the outside of the legs on either side.

In other areas where farmers cannot set posts firmly in thin topsoil they support their fences by placing ingenious pillars along the fence line at regular intervals. In parts of the Missouri Ozarks they have formed hollow cylinders by placing circles of wooden staves inside metal hoops, and they fill these cylinders with loose stones. In the Osage Hills of northeast-

Fig. 9.9.   A wooden crib filled with stones to anchor a fence in a thin-soiled lava area in Oregon.

Fig. 9.10.   A fence anchored by four old tractor tires filled with stones.

ern Oklahoma the cylinders are formed of woven wire. On thin soils de-rived from lava in eastern Oregon, some farmers have nailed together two-by-fours to make sturdy wooden cribs they can fill with stones (fig. 9.9), and at least one clever fellow has formed pillars by filling stacks of four old, worn-out tractor tires with stones (fig. 9.10).

Gates

Farmers who put up fences around fields and pastures must include some way of getting in and out, such as gaps, stiles, gates, and cattle guards. Gaps are the cheapest and most primitive way of opening and closing a barbed-wire fence. The farmer cuts the strands of the fence, attaches the ends to an upright pole, and inserts the ends of the pole into wire loops at the top and bottom of the next post. It is easy to remove the pole from the loops and drag it to one side to allow cattle or horses through the fence.

A stile is a set of steps or a ladder that lets people climb over a fence without harming themselves or the fence. It might be a single step or a set of steps built into a stone wall, or it might be a wooden ladder (fig. 9.11). Farmers hate to see hunters (or anyone else) using the wires of their fences as rungs to clamber over them, because the weight of each person loosens the nails or staples that hold the wire to the posts. Stiles are common in England, especially where popular rights of way or hiking paths cut across individual fields or properties, but in North America traffic is rarely heavy enough to require or justify them.

Gates come in all styles, sizes, shapes, materials, and price ranges to suit the particular needs of each individual farmer. A cattle guard is a set of parallel bars or pipes set at right angles to a driveway to eliminate the need for a gate (fig. 9.12). The bars are set close enough together to support the weight of a car or truck, but too far apart for the hooves of animals. A Y-shaped kissing gate permits two-legged animals to pass, but blocks those that go on all fours. On land subject to floods a conventional fence that traps debris can become an involuntary dam that will eventually be swept away by the water impounded behind it, and the farmers must interrupt their fences with wooden floodgates that will float upward with the rising waters and then settle back into place once the waters have subsided.

Corrals and Feedlots

Ranchers need corrals where they can round up their cattle for veteri-nary treatment, branding, and loading; corrals often are subdivided by

Fig. 9.11. A wooden stile enables people to climb over a barbed-wire fence without damaging the fence.

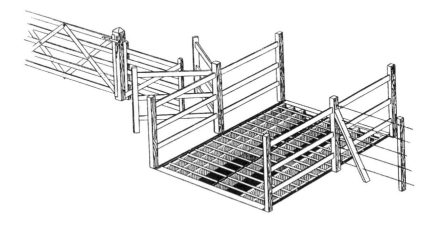

Fig. 9.12. A Y-shaped kissing gate and a cattle grid. Reproduced from *Cattle Grids for Farm and Estate Roads*, Fixed Equipment of the Farm, leaflet no. 7, Ministry of Agriculture and Fisheries (London: H.M.S.O., 1952). Crown copyright is reproduced with the permission of the Controller of Her Majesty's Stationery Office.

cross fences into smaller pens where groups of animals can be separated for special handling. Farmers who fatten cattle need feedlots with troughs into which feed can be unloaded from a slowly moving vehicle along one entire side. Cattle get nervous when they are penned up, so corrals and feedlots must have exceptionally sturdy fences, commonly of boards nailed to massive four-by-four wooden posts or old railroad ties (fig. 9.13).

Corrals and small feedlots usually are set close to the house to enable the owner to keep an eye on the animals, but some commercial feedlots in the West are enormous, with capacities of fifty thousand head of cattle or more (fig. 9.14). The cattle have plenty of drinking water, but feedlots often are called dry lots because they have no grass or other greenery. Stoutly fenced feedlots are becoming common on dairy farms in the East because farmers have learned that it is more efficient to bring the feed to the animals than to turn them out to pasture.

UNUSUAL AMERICAN FENCES

Snow Fences

Snow fences are made of wooden slats wired loosely together with open spaces between the slats. Once they were a common sight on the winter landscape of the snowbelt states. In Minnesota alone, highway departments used to unroll and erect more than a thousand miles of snow fences each fall, and then roll them up again each spring. The fences were strategically placed to force the wind-driven snow to drift behind them instead of blocking highways. They still are used in difficult areas, but they have become less necessary in Minnesota because roads have been rebuilt to eliminate deep cuts where snow could accumulate, they are built higher with wider ditches, and modern snowplows can throw the snow farther away from the highways. Snow fences are still common, however, in other parts of the country.

Stump Fences

Stump fences have virtually disappeared from the cutover lands on the sandy glacial outwash plains of Wisconsin, Michigan, and Ontario. Lumbermen cut down the magnificent white-pine forests that covered these areas, and then they moved on, leaving vast wastelands speckled with stumps that farmers had to pull out before they could cultivate the land.

Fig. 9.13. Beef cattle in a stoutly enclosed feedlot on a Corn Belt farm.

The stumps had tough root systems that were widespreading but shallow, and after the farmers had pulled them out of the ground they dragged the stumps to the edge of the field, tipped them over onto their sides, and lined them up to make fences that were quite effective (fig. 9.15).

Stump fences, like all other wooden fences, are vulnerable to fire and to decay. Outsiders think they are picturesque and like them, but most farmers think they look so primitive and backward that they have replaced them just as soon as they could afford to do so. Those that remain have become so overgrown with weeds and brush that they look like hedgerows, and only close inspection will reveal what is left of the original roots and stumps.

Stone Walls

In 1871 a national census of fences found one-quarter of a million miles of stone walls, enough to go around the world ten times, in the glaciated hills of New England and New York. A few of these structures definitely are well-built walls, but most are no more than haphazard heaps of stones that merely line the edge of a field and do not enclose anything. These pictur-

Fig. 9.14.   Beef cattle on a huge commercial feedlot.

esque piles of rocks may be called stone walls by those who see what they have been told to see, but anyone familiar with proper stone walls suspects that many of these so-called walls are little more than waste heaps of stones dragged from the fields.

In 1988 a wealthy gentleman farmer in Connecticut became angry when his neighbors took some of the stones from one of his "walls" to build their new house, and he invoked a 1902 law that calls for the town selectmen to become "fence viewers" to mediate such disputes. The select-men, after carefully viewing all the walls on his estate, concluded that they were no more than piles of stones, and they ruled that the culprits, who ad-mitted that they had taken the stones, were obligated only to replace them with new stones.

Farmers in New England will tell you that it is positively sinful to walk across a plowed field without carrying a stone in each hand, because each winter's frosts heave a new crop of stones to the surface, replacing those that the farmer removed the preceding summer. Frost heave is a natural process. The soil expands when it freezes, and each time it expands it lifts even the largest stones ever so slightly, until eventually they appear at the surface. The depth of frosts and thus the rate of frost heave are affected by

Fig. 9.15.   A stump fence on a sandy outwash plain.

how well the ground is insulated by trees, leaves, and snow. The early New England farmers may have helped increase the stoniness of their soil when they cut down the forests. The destruction of the natural insulation of the soil could have allowed frosts to penetrate more deeply and thus heave greater numbers of stones to the surface.

The Bluegrass

The gently rolling limestone plain of the Inner Bluegrass basin around Lexington, Kentucky, is a veritable paradise for anyone who delights in fences, because few other areas in North America are so completely fenced and few have such an attractive variety. Within a fifteen-mile semicircle north of Lexington are more than one hundred Thoroughbred racehorse breeding farms, whose owners must be able to spend a thousand dollars as casually as most of us would spend a dime. The hallmark of the Bluegrass horse farm is its wooden fences of three or four horizontal boards that enclose pastures, paddocks, and training tracks. Once the boards were painted a dazzling white, but now they are more likely to be painted black or treated with a preservative to make them last longer.

On horse farms, board fences are practical as well as picturesque, be-

Fig. 9.16.   A double fence on a horse farm in the Kentucky Bluegrass area.

cause the high-spirited horses could easily injure themselves by galloping heedlessly into a wire fence. Even woven-wire fences are topped with a single horizontal board that the horses can see, allowing them to avoid the wire beneath it. Frequently an outer stone wall is paralleled by a board fence a foot or two inside it to protect the valuable horses from being damaged by the stone wall (fig. 9.16). Although board fences are functional on horse farms, they have become so prestigious that they have been adopted as pretentious affectations on beef-cattle farms, on dairy farms, and especially on residential farms.

The early settlers of the Bluegrass Region built their fences of wooden rails, as was customary on the frontier, but by 1840, when timber for rails was starting to become scarce, landowners began to build "rock fences" of limestone quarried from the ledges that outcropped in abundance on their farms. Conventional wisdom, which so often is no more than ignorance, cherishes the myth that black slaves built these rock fences during the winter months when there was little else to do on the farm, but Carolyn Murray-Wooley and Karl Raitz have exploded that myth by showing that the early rock fences were built by specialist stonemasons who brought their skill directly from Scotland or from Scotland by way of Ireland.

Considerable skill was necessary to build unmortared rock fences five

Fig. 9.17. The litter of stones from a disintegrating rock fence attests to the skill of stone-masons who prided themselves on picking up each stone only once, and constructed dry-stone walls that have lasted for more than a century.

Fig. 9.18. Farmers in the Corn Belt have "pulled" their fences when they have switched from mixed crop-and-livestock farming to cash-grain farming.

feet high with smooth sides that tapered from a width of three feet at the base to eighteen inches at the top. The stonemasons used wedges and sledgehammers to crack off uniform, flat limestone slabs two to six inches thick from hillside quarries that were as close as possible to the fence sites. They dry-laid the slabs without mortar in level courses, and capped the fence with triangular slabs set on edge. Each slab in the fence had to fit firmly on half of each of the two slabs beneath it to make a tight joint. Today's beautifully weathered and moss-grown rock fences attest to the skill of the stonemasons who built them more than a hundred years ago (fig. 9.17).

By the 1880s, fences of quarried limestone lined fields, farms, and both sides of nearly every rural road in the Inner Bluegrass, but labor had become so expensive that only wealthy people could afford to build such fences, and they came to symbolize wealth and social status, like the white board fences that replaced them when railroads began to bring cheap lumber from distant areas. Many roads were treacherously narrow for automobiles, and during the 1920s and 1930s stone walls were systematically demolished to permit the roads to be widened. The stones were crushed and used for road construction.

FENCES AND THE FARM ECONOMY

In 1953 Eugene Mather and I explored the relationship between fences and the farm economy along two eight-hundred-mile traverses between Athens, Georgia, and Cleveland, Ohio. In the Middle West, as in other, better mixed crop-and-livestock farming areas in the United States, the standard farm fence was woven wire on wooden posts topped by a strand or two of barbed wire. Insofar as possible the fields of each farm were uniform, interchangeable units, squares of ten to forty acres. Each field had to be securely fenced, because in any year of the standard crop rotation the farmer might use it for rotation pasture, or might turn hogs onto it to "hog down" the standing corn or to glean the fallen corn that the picker had missed.

In 1953 some farmers in the Corn Belt had already shifted from mixed farming to the specialized production of grain crops, primarily corn and soybeans, for direct cash sale from the farm. They had pulled out their fences, which they no longer needed because they had no livestock (fig. 9.18), and they had lengthened the fields to reduce the time they wasted turning large machines around at the ends of rows of crops.

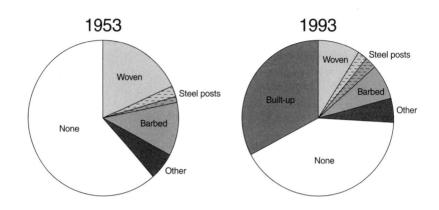

Fig. 9.19. Types of fence along a traverse between Georgia and Ohio. The types of fence did not change much between 1953 and 1993, but one-third of the rural roadsides that had been farmland in 1953 were built up by 1993.

Fences were sporadic in the hills of Appalachia, where the topography is rough, the soils are stony and infertile, and much of the land is steep and wooded. The cultivated land, which produced corn and hay for cattle feed, was concentrated on the strips of bottomland along the streams. Little cropland was fenced, because the cattle were fed in barns during the winter and turned onto hillside pastures from May to November. The hillsides, some surprisingly steep, had originally been cleared for cornfields, but they had already lapsed into unimproved pastures of low carrying capacity that were fenced haphazardly with barbed wire, and many showed signs of reversion to brush and woodland. Some of the more isolated hollows still had a few relict split-rail fences and inhabited log cabins.

The Piedmont area of Georgia and the Carolinas was still recuperating from the demise of King Cotton. Cotton patches had needed no fences, and the only enclosures had been scrawny mule pastures near the farmsteads. Some landowners were trying to convert their former cotton lands into improved pastures for beef cattle by seeding and fertilizing them, and by fencing them with three strands of barbed wire on flimsy wooden posts. Few farmers with small holdings could afford the capital improvements necessary to shift to beef cattle, however, and broom sedge, pine seedlings, and other quick colonizers were rapidly invading many old cotton fields. In 1953 the Piedmont was a wonderful place to observe plant succession on abandoned cropland.

In 1993, forty years almost to the day after our original survey, I replicated our traverses to find how fences had changed (fig. 9.19). My hopes of matching study sites and photographs were dashed by four decades of highway rerouting, straightening, and improvement that had greatly modified many roadsides and opened once-remote rural areas to long-distance commuting. Twelve of our original thirty-two rural survey sites had become so built up that I had to substitute other sites that were as closely comparable as possible, but nonetheless I found that fences have changed to reflect changes in the rural economy.

Fences are disappearing in the Corn Belt, where the shift from mixed crop-and-livestock farming to cash-grain farming is virtually complete. Few farmers still have hogs or cattle, and they no longer need the fences that once enclosed these animals (fig. 9.20). Farmers have removed most of their fences, and those that remain have blossomed into unpremeditated hedgerows overgrown with bushes, briers, saplings, morning-glory and honeysuckle vines, and weeds. Often I had to look carefully to spot the derelict fence beneath the tangle. These overgrown fences still stand because it is simply too much trouble to get rid of them, but they serve little purpose unless they mark property lines.

In contrast to the Corn Belt, the Piedmont and some of the less disadvantaged parts of Appalachia actually had a few more fences in 1993 than in 1953. Many of the steeper hillside pastures in Appalachia have completely reverted to woods that neither need nor warrant fencing, but some landowners have developed extensive cattle-rearing "ranches" in less difficult areas that still are not well suited to crop cultivation. These new ranches have only cattle and horses, so their standard fence is three or five strands of barbed wire, often on steel posts, but affluent people may invest in woven wire even though they do not really need it.

Refugees from the city have adopted types of fence that few farmers can afford anymore. Some of the new cattle ranches have been developed by well-to-do city people who use white board or post-and-rail fences to advertise their status. Less affluent people have bought a few acres out in the country where they can keep a horse for the children; they publicize their ownership by erecting white board fences. White board fences symbolize horses and money, and post-and-rail fences give an aura of age to new suburban properties, but wire fences on the other three sides of an enclosure may expose the sham fences along the roadside.

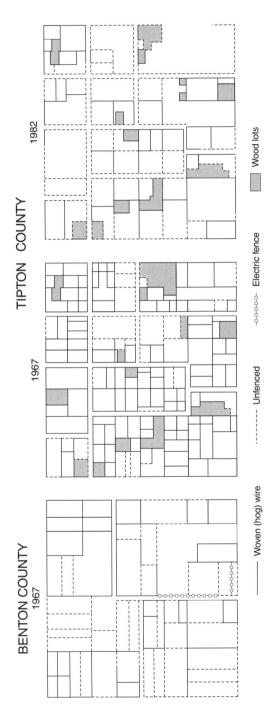
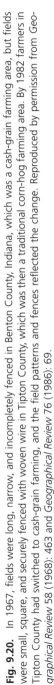

Fig. 9.20. In 1967, fields were long, narrow, and incompletely fenced in Benton County, Indiana, which was a cash-grain farming area, but fields were small, square, and securely fenced with woven wire in Tipton County, which was then a traditional corn-hog farming area. By 1982 farmers in Tipton County had switched to cash-grain farming, and the field patterns and fences reflected the change. Reproduced by permission from *Geographical Review* 58 (1968): 463 and *Geographical Review* 76 (1986): 69.

🦋 A New Settlement System

One of the unanticipated results of my resurvey of farm fences was a growing awareness of the extent to which residences have preempted rural roadsides, especially in the southern manufacturing belt of the Carolinas, Georgia, and eastern Tennessee, where a distinctive new settlement system has evolved. In 1953 we had restricted our attention to farm fences, but at least a quarter of the roadsides that had been farmland in 1953 had been built up with rural residences by 1993, and the remaining farmland had been cordoned off by residential development along the highways (see fig. 9.19). At some of our original survey sites I had difficulty finding any roadside farmland at all.

In much of the eastern South, especially in the manufacturing belt, one is rarely out of sight of a rural nonfarm residence on any stretch of paved highway (fig. 9.21). These residences are a heterogeneous mixture of weatherbeaten old shacks, immobilized mobile homes, suburban-style houses, and costly mansions, with a farmstead only every now and then. The pattern is so sporadic, so unorganized and unplanned, that "bead" development describes it better than the more usual "ribbon" or "strip" development. It has become ubiquitous, and has evolved so gradually that most people take it for granted and few have even bothered to comment on it. It is interrupted only where the land is in large holdings and strong hands, where it belongs to people who can afford not to sell off plots of land along the highway.

Immobilized mobile homes have become the residences of choice, or at least the starter homes, for many people. Some are segregated in trailer parks but most are scattered along the highways and byways on individual plots of ground. In Appalachia mobile homes have completely replaced log cabins; the only log cabins that still are occupied are those that have been taken over by yuppies and gentrified. There still are shacks aplenty, but there are also many attractive new houses with spacious lawns maintained like golf-course fairways, and late-model cars parked in their driveways.

Some of these roadside residences are occupied by farm people and their children who have migrated to the blacktop, but most are the homes of city people who have moved to the countryside. Long-distance commuting has become the new lifestyle in an area where winter driving holds few terrors. At convenient spots along the highways there are formal and informal parking lots for carpools, where people can leave their cars for the day and make the long drive to work in a single vehicle.

There are no remote and isolated areas left in the South, not even in

Fig. 9.21. One is rarely out of sight of a rural nonfarm residence on any stretch of paved highway in Spersopolis, which includes the Piedmont of the Carolinas and Georgia and the Great Valley of Upper East Tennessee and southwestern Virginia.

Appalachia. The construction of the Interstate Highway System has been given credit—perhaps too much credit, because the individual states have capitalized on the know-how developed in constructing the interstates and put it to good use in upgrading their own highway networks. Most federal and state highways are good to excellent, and it is easy to commute from anywhere to one of the new factories that have been established in the area. Traffic is heavy when shifts are changing, but it moves rapidly. Only the last few miles between home and the carpool parking lot on the highway are in poor shape, and they carry little traffic.

At first blush it might seem strange to wrap up a discussion of fences and fields with a description of a new settlement system, but I have done so because it so clearly demonstrates the interrelatedness of all geography, and the importance of keeping your eyes open. My search for farm fences along rural roadsides made me aware of a distinctive new settlement pattern that has evolved spontaneously and dramatically changed the face of the industrialized South, yet it has evolved so gradually that most people simply take it for granted and fail to realize how distinctive and widespread it is.

10 *Barns*

Barns are the largest, the most imposing, and probably the most misunderstood structures on farms. Most Americans have glanced at barns from the highway, but few have actually ever set foot in one, and fewer still have any sense of their role in the farm economy. Barns are farm implements, every bit as much as plows and tractors and combines and fences. Farmers have built barns because they have needed places to store their crops and to shelter and feed their animals, and each barn reflects the needs of the farmer who built it. An appreciation of barns must begin with an understanding of their function, the purposes they were designed to serve, the reasons why they were built. We must learn to see them through the eyes of the farmers who have built them and paid for them, who have used them and modified them, who have maintained them or allowed them to fall into disuse.

Building a barn is one of the greatest single financial outlays a farmer will make in a lifetime, and most farmers think long and hard and do a lot of figuring before they finally decide to invest in building one. Few farmers command the technical competence necessary to build a robust barn, and most seek the skills of a master carpenter or barn builder, at least to raise the frame, once they have decided that they need a barn and can afford to pay for it.

They are hostage to local barn builders. For example, my uncle Ashby was one of only two barn builders in our community. He knew how to build only straight gable roofs, whereas the other fellow built only gambrel roofs. If you got Ashby mad at you (which required no great skill, because he was a cantankerous old cuss), you wound up with a gambrel roof on your barn whether you wanted it or not, because that was the other fellow's trademark, and it was the only kind he would build.

To the untutored eye each barn might look unique, and at first glance the variety of barns, even in a small area, can be bewildering. Beneath their details of whim and trim, however, the great majority of barns in any region are generally similar because most farmers in the region have similar needs, and they have built similar barns to meet them. Barns differ from region to region, however, because the farmers in different regions have different needs, and thus it is possible to identify regional types of barn that are associated with distinctive regional types of farming.

An understanding of barns requires an understanding of the past as well as the present, of what farming used to be like and how it has changed, because the barns of a region become obsolete when its farming system changes. New types of barn must be developed to serve the new farming system, but some old barns still stand. Farming has changed so much in recent years that any farm building more than about thirty years old probably is obsolete, because it has outlived the function it was originally built to perform. Barns are no exception, and many old barns have been modified or razed.

BARNS OF WESTERN EUROPE

In western Europe barns are always and only associated with grain farming. The word *barn* is derived from the Old English word *bereaern*, which means "barley place" (from *bere*, barley, plus *aern*, place), and it can be used only for the farm building in which grain is threshed. The grain might have been stored in the barn after it had been threshed, but many farms had granaries, separate buildings for storing grain, and they used their barns only for threshing.

Barns originated in northwestern Europe to shelter sheaves of grain and threshing floors. In the mild, dry winters of the lands around the Mediterranean Sea the sheaves could safely be left outdoors and the grain could be threshed in the open, and the early Spanish missions in California used open-air threshing floors (see fig. 8.2). In the wetter and cooler areas north of the Alps, however, the sheaves had to be stored under cover until they were threshed because the grain would rot or sprout if they were left out in the fields, and the workers needed protection from the elements while they were threshing the grain.

In western Europe barns are never used to house livestock. It is unthinkable that animals should ever be allowed in the barn, and each type of farm animal had its own structure with its own name. In England, for example, any country person can tell you that the barn is only for threshing; horses belong in the *stable*, cows in the *byre* or *shippon*, and pigs in the *sty*. In France you thresh in the *grange*, but keep horses in the *écurie*, cows in the *étable*, and porkers in the *porcherie*. In Germany, as nearly as I can tell (although lexicographers, urban types all, are not very helpful when it comes to making such distinctions), you thresh in the *scheune*, and keep horses in the *pferdestall* and cows in the *kuhstall*. The names *shippon* and *byre* failed to make it across the Atlantic. Even though *stable* and *sty* successfully weathered the crossing, in American usage the meaning of the word *barn* has been expanded to include almost any big, old building on a farm, and a barn may be used to house crops, livestock, machinery, or almost anything else one can think of.

The modular barn of western Europe in bygone years was a simple rectangular structure, high as a two-story house, with large double doors opposite each other in the two long sides (fig. 10.1). One door opened into the rickyard and the other into the stockyard (figs. 10.2 and 10.3). At harvest time the sheaves of ripened grain were built into neat, thatch-covered ricks in the rickyard, or stacked in the ends of the barn. The grain was threshed with hand flails on the barn floor between the two doors, which were opened to create a draft to blow away the dust and chaff. Right through the winter two workers flailed away on the threshing floor each day until they had produced enough straw to feed and bed the cattle in the stockyard. Straw is the hollow stems of grain plants, principally wheat or oats, after threshing has removed their grain heads; it has limited nutritional value, but is useful for bedding. The grain not needed for feed was stored in the ends of the barn, which might or might not have lofts. The stockyard was the farm fertilizer factory; although the farmer might hope to sell the fattened animals at a profit, and often did, their principal contribution to the farm was the manure they produced.

On many farmsteads the basic module of rickyard, barn, and stockyard is obscured by a clutter of other structures, especially if the farmstead is jammed in amongst the other buildings of a village, as so many are. The module can be seen most easily in the field barns that have been built on isolated parts of awkwardly shaped farms (see fig. 7.10). Some of the farms created by enclosure, for example, were shaped like slices of pie, with their apexes at the old farm buildings in the village. The widest part of such a farm, at the back end, might be as much as a mile or more from the farmstead. In order to avoid having to haul sheaves of grain all the way to the farmstead and then having to cart the manure back to the fields, the farmer built a field barn at the back end of the farm. Field barn complexes con-

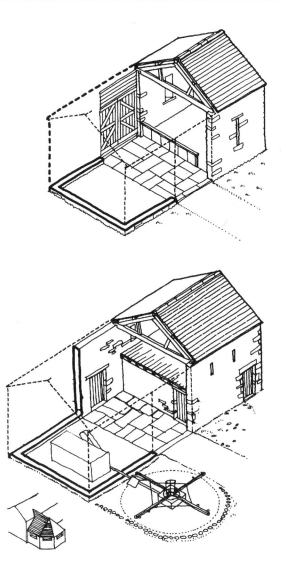

Fig. 10.1. Cutaway isometric drawings of three-bay barns in England. Workers threshed grain with hand flails on the central threshing floor, for which large opposing double doors provided light and a draft to blow away the chaff. The bottom drawing shows how three-bay barns were modified when threshing machines replaced hand flails. The threshing machine was placed in one bay, and it might be powered by a waterwheel, a windmill, a steam engine, or most commonly, as in this sketch, by horses plodding in circles. The power source was housed in a distinctive structure beside the bay of the barn that held the threshing machine. Both drawings are reproduced by permission from a splendid book by R. W. Brunskill, *Illustrated Handbook of Vernacular Architecture* (London: Faber and Faber, 1971), 141 and 143.

sisting of a rickyard, a barn, and a stockyard are common features of the rural landscape in western Europe, and they may be seen on many detailed topographic maps.

Most barns in Europe are more than a century old, and many were extensively modified when threshing machines replaced hand flails. Some farmers permanently installed new threshing machines in the ends of the old barns because that was where grain had always been threshed. The threshing machines required power, and in the south of Scotland at least four different sources of power are indicated on detailed topographic maps by otherwise mysterious projections on the back sides of barns. A long, narrow projection paralleling the long side of the barn probably housed an undershot waterwheel, driven by a small stream whose diverted waters were stored behind a dam in a tiny farm millpond. A small block projection might have contained a coal-fired furnace for steam power (fig. 10.4), if it had a tall smokestack, or it might have been the base of an old windmill (fig. 10.5). A round or octagonal structure probably sheltered a treadmill to which a horse was hitched; the threshing machine literally was operated by one horsepower (see fig. 10.1).

EUROPEAN BARNS AND FARMSTEADS

The early immigrants to North America from western Europe were each familiar with one of three patterns of barn placement on farmsteads. The farmsteads of the first type had no barns because they were in areas so damp and so cool that oats was the only grain crop feasible, and even that crop was rarely planted. People and their animals were sheltered under the same roof in a single small structure, the people in one end and the animals in the other. The crop of oats, if it ripened, was so meager that it could be threshed on the kitchen floor, with the front and back doors opened so the chaff could blow away. Examples of this type of structure, such as the *maison-bloc* of France, the long house of Wales, and the traditional peasant cottages of the wetter parts of Scotland and Ireland, were only in the poorest and most primitive areas, and few remain today, although the birthplace of Robert Burns, the national poet of Scotland, has been restored to its original condition (see fig. 7.3).

Farmsteads of the second type also consisted of a single structure, with people, crops, and animals all under the same roof, but the structure was much larger than the peasant cottage, with far more internal differentiation

Fig. 10.2. The rickyard side of a wooden barn with a thatched roof in southern England. The large doorway leads to the threshing floor where the grain was threshed. (Cf. fig. 7.10.)

Fig. 10.3. The stockyard side of the barn in fig. 10.2. In the stockyard the farmer fed beef cattle to produce meat and manure.

(fig. 10.6). The living quarters, commonly several stories high and one room deep, were in one end of the building. A wide transverse central passage separated them from the livestock area in the other end. The main entryway was through large doors in either side of the structure that opened onto this passage, or "threshold," where grain was threshed in season. This type of structure was most common in the northwestern marshlands and in the southern hills and mountains. Saxon invaders from the marshlands took the idea with them to England, but the various functions eventually were moved to separate buildings.

The farmsteads of the third type have separate buildings, including separate barns, for separate farm functions. In England and in central Sweden these buildings are scattered around the farmstead with no apparent order or plan, but in other areas the buildings are arranged around a central courtyard, usually in some formal, standard order. In Denmark, for example, many farmsteads have the farmhouse facing the barn across the courtyard, with the stables on the left and the byre on the right as you walk out the back door of the farmhouse.

BARNS OF NORTH AMERICA

Farmers build barns because they need them. Why did farmers in the various parts of North America need barns, and how have they used them? Barns can best be understood in context, as integral components of the farming systems they were built to serve.

Construction

Some people are fascinated by the methods and techniques of barn construction, and they delight in exploring complex structural systems of beams, braces, posts, rafters, purlins, sills, plates, struts, and so forth. No one wants a barn to be blown away by the first puff of wind that comes along, and most farmers will use the best technology available when they build their barns, but details of construction are not necessarily related to the ways in which barns are used.

Some of the most obvious features of barns are equally unrelated to their use. Barn roofs, for example, are not nearly so significant as they might seem, although they do suggest when the barns were built. Straight gable roofs are the easiest to build and the oldest, but they are also the least effi-

Fig. 10.4. An English three-bay barn that has been modified by the placement of a thresh-ing machine in the far bay. The steam engine that powers the threshing machine is in the structure on the left, which has a tall smokestack to ensure a good draft.

Fig. 10.5. An English three-bay barn that has been modified by the placement of a thresh-ing machine in one bay. The tower once housed a windmill for power, and the small round structure housed the gin (engine) house, where plodding horses provided power on days when the wind was not strong enough to drive the windmill (see fig. 10.1).

cient because you are in constant danger of hitting your head on the rafters when you are working under the eaves on either side. Gambrel roofs have more headroom under the eaves, but building them requires more skill, and in most areas they did not become common until after around 1880, or even later, when lumberyards started stocking prefabricated gambrel trusses. Around 1920 lumberyards began to stock arched Gothic trusses, and most barns built since then have had arched Gothic roofs, but few barns have been built since then. Some old barns have been reroofed with newer gambrel or Gothic roofs, but the newer types of roof are most common on extensions that have been added to older barns (fig. 10.7).

Distinctive building materials and colors certainly contribute to the picturesque character of barns, but the use of local materials should not be unduly romanticized. Farmers use local materials because they are cheapest, not because the farmers have some mystical attachment to the land, and they could just as easily render the same basic barn in log, in brick, in stone, in board, or even in stovewood. The color of a barn, whether red, white, or unpainted wood weathered to a silvery gray, might reflect the affluence or preference of the farmer, but, like the building material, it does not affect the way the barn is used.

The size and placement of doors and other means of access, which reveal how a barn could be used, are more significant than its roof, building material, or color. Small doors can admit people and animals, but wagons and machinery require larger entryways. The principal doors of threshing barns usually are in the sides of the barns, but the principal door of a hay barn often is in the end. A barn with two levels must have some means of access to the upper level—an earthen ramp, or barn bank, up which the farmer can drive a loaded wagon, or an opening for a hayfork high in the gable end. The upper level of two-level barns is for threshing or hay storage and the lower level houses animals. Individual farmers have their own particular ideas about the arrangement of the animal areas in their barns, and it is unusual to find two barns with the same internal layout on the ground floor.

The size and height of barns, their roofs, their building materials and color, their entryways, their ground plans (rectangular, square, ell, round) and internal layout, and their appendages, such as silos and feedlots, may reflect their dates of construction, the specific needs and whims of individual farmers, and the size and complexity of their farm operations. Those who make detailed studies of the individual barns in any small area can become mesmerized by their diversity, and they may lose sight of the basic

Fig. 10.6. Saxon farmsteads had all farm buildings under a single roof, with the living quarters on the right, large double doors leading to the threshing floor in the center, and stalls for livestock at the back. The idea of living under the same roof with livestock was unpopular in England, and it never caught on in North America.

Fig. 10.7. The arched roof of the addition on the right says that it is newer than the barn, which has a gambrel roof.

functional similarities that identify the major regional types. Local studies of barns are desirable and necessary, but they must be placed in the context of the larger regional patterns of order that underlie this local variety.

Ethnicity

Most attempts to relate the barns of the North American Midwest to European ethnic groups have been unsatisfactory. It is tempting, but unwise, to identify an area that was occupied by a particular ethnic group and to proclaim that the features of the area are unique and distinctive to that group, even though these features differ not at all from the features of adjacent areas occupied by different groups.

Few settlers migrated directly from the old country to the North American frontier until the 1850s, when the construction of railroads permitted long-distance travel to the port of departure in Europe, and from the port of arrival to the frontier in North America. The prerailroad frontier east of the Mississippi River was settled in a series of short leapfrog moves by young people from areas that had been settled earlier, not by immigrants. For example, the "English" people who settled Michigan and Wisconsin were from New York and New England, not from England, and the "Germans" who settled southwestern Ohio and Indiana came from Pennsylvania, not from Germany. The European ideas had been modified by the North American experience.

After the 1850s, when settlers could migrate directly from Europe to the frontier, they found Americans practicing an ecologically sound farming system that had prospered economically for half a century or more, and the newcomers had sense enough to adopt it hook, line, and sinker. They might put a few superficial touches, such as decorations, on their barns and farmsteads to remind them of the old country, but their basic barns and farming systems were just like the barns and farming systems of their neighbors.

BARN TYPES AND FARMING SYSTEMS

Broad regional patterns of barn types have developed in the eastern United States in close functional association with major agricultural systems. The barns on dairy farms in Wisconsin, for example, are quite unlike the livestock feeder barns on Corn Belt farms in Iowa, and farms in the hard hills of Appalachia have small structures that farmers from the Midwest scorn as mere sheds rather than proper barns. The plains of the South, former fiefdom of King Cotton, have had few barns because the traditional crops of the region—rice, peanuts, and sugarcane as well as cotton—did not require them, and farmers needed only sheds for their work animals, or workstock. Tobacco, the other major southern crop, is a completely different story, however, because each tobacco district requires its own distinctive structures for curing the leaves. Other specialized crops, such as potatoes and hops, also require distinctive structures that might or might not properly be called barns.

Farmers have had to develop new types of barn when farming systems have changed, whether geographically, when settlement has spread into areas with a different climate or topography, or historically, in response to changes in agricultural technology or markets. Five key areas have been especially important in the development and spread of barn types in the eastern United States (fig. 10.8). Ideas from Europe were transplanted to southern New England and southeastern Pennsylvania. Farmers developed new types of barn in upstate New York and in Upper East Tennessee in response to major changes in agriculture. And two major barn-type concepts collided in southwestern Ohio, with one emerging victorious to become the standard barn type of the Corn Belt.

New England and New France

The early settlers in New France (Quebec) and New England built three-bay barns like those they had known in Europe (fig. 10.9). (The major structural divisions of a barn are called bays.) A three-bay barn is a hollow, rectangular shell two stories high with large double doors in both sides leading to a central threshing floor with bays for grain storage in either end. These barns were single-purpose structures in which wheat and other grains were threshed and stored. They were barns, not stables or byres; the very idea of keeping horses or cattle in one of them would have made about as much sense to their owners as the idea of keeping horses or cattle in their parlors or kitchens.

The early settlers continued the European tradition of erecting separate structures for each farm function. A farm might have a variety of small sheds for cows, oxen, horses, pigs, sheep, chickens, hay, corn, sugaring, wagons, and carriages, or it might have none. Farmers who did not have hay sheds simply stacked their hay in the fields until they needed it, and those who

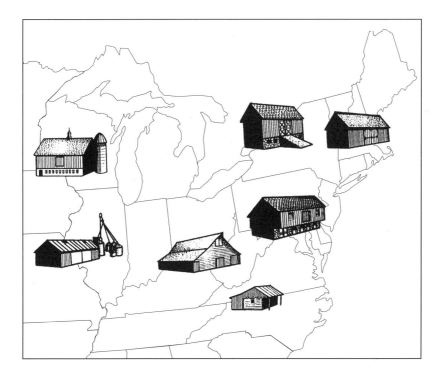

Fig. 10.8.   Key areas in the development and spread of barn types in the eastern United States: three-bay barns in New England, raised three-bay barns in upstate New York, Pennsylvania German forebay barns, single-crib log hay barns with side sheds in the Great Valley of southwestern Virginia and Upper East Tennessee, Corn Belt livestock feeder barns, Wisconsin dairy barns, and contemporary metal machine sheds flanked by metal grain bins and an elevator leg. Michelle Terrell drew the original map, which was first published in *Focus* 43, no. 1 (1993): 8, and is reproduced by permission.

left their cattle outside all winter apparently were unbothered by routine criticism from visitors. In Quebec some farmers attached cowsheds to the ends of their three-bay barns (fig. 10.10). The windows and small doors of the cowsheds contrast sharply with the wagon-wide double doors of the barn, and they show that the later additions are separate structures even though they are under a single roof.

Eventually the settlers in New England began to modify their ideas about barns when they adjusted their farming systems to the climate. Wheat did not ripen well in the cool, damp summers, and hay became a more important crop. Hay consists of grasses or other plants, such as clover and alfalfa, that have been mowed and dried for use as winter feed for livestock.

Fig. 10.9.   A log three-bay barn in the province of Quebec.

Farmers had to make hay to feed their animals from the time the pastures dried up at the end of the summer until the grass began to grow again in the spring. A mature horse or cow needs about two tons of loose hay, or roughly a ten-foot cube, to see it through a winter. Both hay and cattle benefited from shelter, so eventually farmers modified their traditional barns (fig. 10.11). They filled one end of the barn completely with hay, built stalls for the animals in the other end, and built lofts for hay storage above the stalls (fig. 10.12). They drove loaded hay wagons onto the former threshing floor and muscled the hay up into the lofts with pitchforks.

Devotees of the Doctrine of Retrospective Inevitability ("it seems perfectly logical to us now, so it must have seemed equally logical to them back then") like to believe that Yankee farmers began to use their three-bay barns for cattle and hay quite early on, but just the opposite is probably the case. It might seem logical to us today, but in a traditional and conservative society it must have seemed downright revolutionary to let cattle into the barn, and the first person courageous enough to do so must have encountered considerable scorn and ridicule from other farmers. "He's keeping cattle in his barn!" they must have sneered.

By 1820 most progressive farmers in New England probably were using their old three-bay barns for cattle and hay, and they were starting to build large new barns like those that farmers were building in other parts of the United States. These new barns had their principal entryways in their ends rather than in their sides, and they were hay barns that were not intended for threshing. Animals were housed under a hayloft on one side of the central driveway, and the other side was for hay.

By 1850 farmers in New England were feeling the pressure of competition from newly settled areas in the Midwest, which could produce anything that New England farmers could produce, and could undersell them even in New England. With the exception of a few favored areas, New England is a marginal agricultural region. The summers are cool and damp, the soils derived from glacial deposits are stony and infertile, and the glacially molded topography is so choppy that many fields are too small even for horse-drawn machinery; some farmers were still using oxen rather than horses for draft animals as late as 1900.

Traditionally, New England farmers had produced small surpluses of wheat, butter, cheese, beef, pork, potatoes, and other farm products for local

Fig. 10.10. Farmers in Quebec expanded their three-bay barns by adding structures to their ends. The barn on the left has been expanded by adding the multiwindowed cowshed on the right. The door painted on the end of the barn is a trompe l'oeil.

Fig. 10.11. A traditional three-bay barn was a hollow rectangular shell with no internal structures.

markets, but few could produce truly commercial quantities. Most farm families had to have nonfarm income, whether from work in the woods as lumbermen in winter or from a wide variety of home processing activities, including butter and cheese making, food canning and preserving, spinning and weaving, quilting, soap and candle making, woodworking and leatherworking, and the making of clothing and shoes.

The prideful Yankee farmers who were still trying to wrest a living from the harsh land of New England after 1850 could not change the environmental constraints under which they labored, and they could not compete with farmers in the Midwest, but they could and did try to rationalize and modernize their farmsteads to make them more efficient and more attractive. They reorganized their existing clutters of small structures by moving them into staggered lines of connected buildings at right angles to the road, thus creating the distinctive connected farmsteads of New England. The idea of moving buildings around a farmstead might seem extraordinary to us today, but people with long experience of moving heavy loads of logs out of the woods on skids drawn by oxen thought little of it.

The farmhouse, or "big house," which was next to the road, had the

Fig. 10.12. Some farmers modified their three-bay barns (see fig. 10.11) by building enclosed stalls for livestock in one end.

parlor and bedrooms, and was primarily a place of rest (fig. 10.13). The "little house" behind it held the kitchen and workrooms, and was the daytime center of family activity. In back of it was the backhouse, often the original three-bay barn that had been converted into workshop and storage areas, and behind it was the new end-opening barn for hay and cattle. New England connected farmsteads were part farm and part small-scale factory. They were created to make farming operations more efficient, but they also reflected the need for home industry to supplement meager farm income.

Raised Three-Bay Barns

The three-bay threshing barn moved west from New England with the wheat frontier, first to upstate New York, then on to the Middle West. Farmers in Michigan were building three-bay barns until well into the twentieth century. Each new wheat frontier had the same history: initial bonanza, then soil depletion, problems with insects and diseases, intense competition from new areas farther west, and eventually a shift to dairy farming. In 1840, for example, upstate New York was the nation's principal wheat-

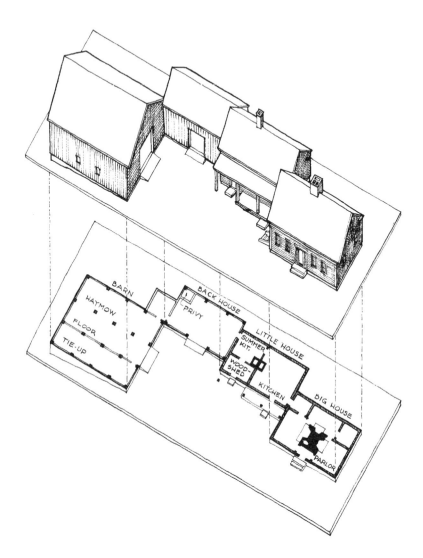

Fig. 10.13. A typical building and room arrangement of connected farms. Reproduced by permission from Thomas C. Hubka, *Big House, Little House, Back House, Barn: The Connected Farm Buildings of New England* (Hanover, N.H.: University Press of New England, 1984), 7.

producing area, but within a decade or so areas farther west had supplanted it, and farmers in New York had begun to concentrate on producing butter and cheese for the cities of the eastern seaboard, to which the Erie Canal and, later, railroads gave them easy access.

High-quality dairy cows needed shelter during the harsh winter months, which are more harsh toward the continental interior, and dairy farmers in upstate New York experimented with different strategies for adapting their three-bay barns to dairy farming. A few, like farmers in Quebec, attached separate cowsheds to the ends of their three-bay barns. Some, especially on small farms, would build a solid masonry cowshed with numerous windows in one end of a three-bay barn, and use the loft above for hay. By 1850 threshing machines had essentially replaced hand flails for threshing, so the threshing floor and the other end of the barn were also used for hay storage. I have dubbed these structures "half-and-half" barns, because I have found no other name for them. A half-and-half barn is essentially a modified three-bay barn.

The half-and-half barn, with a masonry cowshed on one side of a central driveway and a hayloft on the other, became stylized; later, when Wisconsin shifted to dairy farming, many of the first dairy barns on small farms were built from scratch in this fashion. When farmers prospered and needed more stall space for their larger herds, they expanded the cowshed, sometimes by extending it, sometimes by building an addition at right angles, which changed its ground plan from an I-shape to an L-shape.

The most popular strategy for adapting three-bay barns to dairy farming was based on the development of a new and distinctive type of barn by an innovative farmer who got the bright idea of jacking up an entire three-bay barn and building underneath it a masonry ground floor with stalls for dairy cows. The ground floor of this "raised three-bay barn" had numerous windows to admit light and air to the ground floor where the cows were milked and where they spent most of their time during the colder months of the year.

Unlike the traditional three-bay barn, which had only a single level, the raised three-bay barn had two distinct levels, an upper level for hay and a lower level for animals. Farmers needed some means of access to the upper level (fig. 10.14). On flat land they would build a gentle earthen ramp, or barn bank, to allow them to drive a loaded hay wagon onto the upper level, but some farmers excavated part of the ground level out of a hillside, and they could drive the wagon straight onto the upper level from the uphill side of the barn. The term *basement barn* is misleading, however,

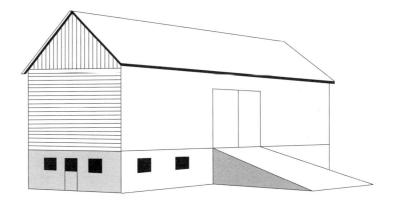

Fig. 10.14. A raised three-bay barn, with a ramp, or barn bank, leading to the three-bay barn on the upper level. Reproduced by permission from Peter M. Ennals, "Nineteenth-Century Barns in Southern Ontario," *Canadian Geographer* 16 (1972): 260.

even though some raised three-bay barns do have partially excavated basements, because raised three-bay barns on flat land have barn banks, and their lower levels are completely above grade.

The raised three-bay barn also became stylized, and it was built from scratch as the standard barn on dairy farms in Wisconsin and Minnesota until the development of hayforks eliminated the need for barn banks. Farmers were delighted to get rid of their barn banks because the lower level beneath them was damp and poorly ventilated. A hayfork dangled from a trolley that ran on a track at the ridge of the roof inside the barn. The farmer used a system of ropes and pulleys to move the trolley back and forth, and to raise and lower the hayfork. The track extended far enough beyond the end of the barn to reach out over a loaded hay wagon parked below. The farmer pulled the trolley to the end of this extension, lowered the fork to pick up a load of hay from the wagon, raised the load, pulled the trolley back to the desired spot inside the barn, and dumped the hay there.

The lower level of a Wisconsin dairy barn with a hayfork was like the lower level of a raised three-bay barn, but the principal entryway to the upper level was moved from the side of the barn to an opening high in the end beneath the extension of the track for the hayfork trolley (fig. 10.15). The roof of the barn was lengthened to cover this extension. The roof extension ranged from an inconspicuous V to an elaborate structure, and it seems to have no vernacular name, but various academics have called it a hay gable, a hay hood, or a hay bonnet. In time hayforks were replaced by conveyors that carried bales of hay into the barn loft, or by blowers that blew chopped hay into the loft through a long metal tube, but the distinctive feature of the new Wisconsin dairy barn remained the opening high in the end.

Nearly every Wisconsin dairy barn also had a towering cylindrical silo beside it (fig. 10.16). The growing season in Wisconsin is so short that farmers feared that corn, their principal crop, would be damaged by frost before the grain had fully ripened. They cut the entire plant while it was still green, chopped it into small pieces, and blew it into silos, where they stored it as winter feed for their cows.

Pennsylvania German Forebay Barns

German-speaking Swiss farmers who settled in southeastern Pennsylvania brought to North America the idea of keeping animals and crops under the same roof (fig. 10.17). A *Schweizerscheuer* (Swiss barn) is a massive two-level structure, eighty to a hundred feet long, fifty to sixty feet wide, and fifty feet to the eaves. On one side a barn bank leads up to wagon-wide double doors that open onto the central threshing floor on the upper level. On the other side the upper level projects three to six feet out over the

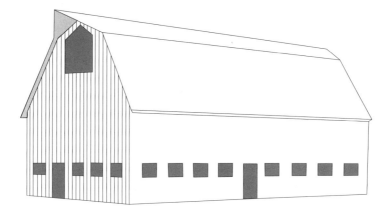

Fig. 10.15. A stylized dairy barn with an opening high in the end through which hay is taken into the loft by the hayfork. The V-shaped roof extension covers the track of the trolley on which the hayfork runs. Modified and reproduced by permission from Peter M. Ennals, "Nineteenth-Century Barns in Southern Ontario," *Canadian Geographer* 16 (1972): 262.

Fig. 10.16.   A Wisconsin dairy barn and silo.

stockyard in a distinctive cantilevered forebay that is supported by large wooden joists (fig. 10.18). This forebay, which is the diagnostic feature of the barn, shelters the entrances to the livestock area on the lower level. These entrances have divided Dutch doors whose top half can be opened even in bad weather to let in air and light, while the bottom half stays closed to keep in the animals and to keep out driving rain and snow.

The projection of the forebay also helped keep the doors clear of straw, which was thrown into the stockyard for feed after the grain had been threshed from it on the upper level. The stockyard on one of these Pennsylvania farms, as in Europe, was the area where the animals produced the manure that the farmers needed to fertilize their fields. Rain and snow quickly leached plant nutrients from manure in the open stockyard, however, and some farmers protected it by building a hay shed at right angles to one end of the barn. The hay shed extended the upper level out over the stockyard and changed the ground plan of the barn from a rectangle to an L-shape.

Most barns in southeastern Pennsylvania have been modified so greatly that it is hard to find an unaltered specimen. Many farmers enclosed the area under the forebays to increase their stall space when they shifted from beef to dairy cattle, and they have added new sheds to the sides and ends of the barns. Most farmers also added one or more silos when they shifted to dairy farming.

Some elements of the Pennsylvania German forebay barn have been copied more widely than others. The idea of a barn bank leading to the upper level might have inspired the development of the raised three-bay barn in upstate New York. The idea that a single general-purpose farm building should have crops and livestock under the same roof was a truly remarkable innovation, but it has been adopted so universally that it is accepted as a self-evident truth by most Americans, and it is hard to convince them that livestock are never allowed in barns in Europe.

Some people have suggested that the idea of two-level barns might have come from a small, isolated part of the remote Lake District in northwestern England. Many Lakeland barns do have barn banks, and some even have pent roofs on the barnyard side (fig. 10.19), but a simple pent roof

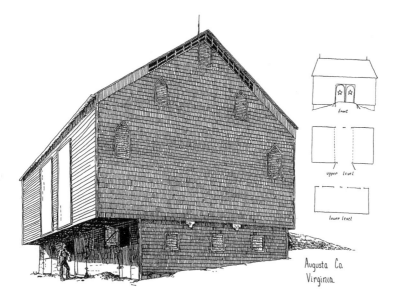

Fig. 10.17. A Pennsylvania German barn with a diagnostic forebay. Reproduced by permission from Henry Glassie, "The Old Barns of Appalachia," *Mountain Life and Work* 40, no. 2 (1965): 27.

Fig. 10.18.   The forebay of a Pennsylvania German barn.

Fig. 10.19.   A two-level barn with a pent roof in the Lake District of northwestern England. Despite superficial similarities, the idea of the Pennsylvania German *Schweizerscheuer* probably was not imported from the English Lake District.

is a far cry from a massive forebay, and it is unlikely that the farmers of southeastern Pennsylvania would have called a Lakeland barn a Schweizer-scheuer, which was the common name for their forebay barns.

The diagnostic trait of Pennsylvania German barns is their forebays, which were never built on other two-level barns. The picturesque barn decorations that have attracted so much attention are localized even within southeastern Pennsylvania. Most of the barns with "hex signs" (so called for the benefit of gullible tourists) are within a twenty-mile radius of Kutztown, between Reading and Allentown; in the 1830s and 1840s, when farmers began to paint their barns, they added these designs to make them look prettier. Brick barns with designs created by leaving selected bricks out of the end walls are mainly south and southwest of Harrisburg. The resulting openings, which admit light and air, perform the same function as the narrow slits and loopholes that were made in the end walls of stone barns. The loft of a barn must be well ventilated because hay stored in a poorly ventilated loft is subject to spontaneous combustion, which can cremate the entire structure.

Cribs and Barns of Appalachia

German settlers brought their knowledge of barn types, a farming system, and techniques of log construction to the United States, but this knowledge was acquired and disseminated by the restless Scotch-Irish, frontierspeople par excellence, who modified it greatly as they moved westward and southward into areas of mixed quality in the Appalachian uplands, where the better farming areas are underlain by limestone. The people who settled on the limestone lands in the Great Valley of Maryland and Virginia continued to raise wheat and build forebay barns, even though the winters were milder and shelter for livestock was not essential, as it was farther north.

Southwest of Lexington, Virginia, however, the Great Valley is severely constricted, and for two hundred miles it is pinched out into cramped and narrow valleys where even mediocre farmland is in painfully short supply. The settlers in southwestern Virginia had to develop a new and different type of farming because the land simply was not good enough to support the Pennsylvania German system. Corn replaced wheat as the staple crop on small patches of cultivated land. It provided food and fiery beverage for the farm family, and feed for a mule or two, a couple of hogs, and a milk cow.

Corn does not have to be threshed. The farmers stored ears of corn in

Fig. 10.20. An Appalachian log corncrib. The simple rectangular log crib was the module from which various barn types were developed in southwestern Virginia and Upper East Tennessee.

simple rectangular log cribs that were smaller, unchinked versions of the log cabins in which the people themselves lived. (A log "pen" is for people, and a log "crib" is for crops and livestock; some academics once allowed themselves to get all hot and bothered about the way the ends of the logs were notched to make them fit together tightly, and they tried to relate types of corner-timbering to ethnic groups, until someone pointed out that often the logs are notched differently even on a single structure.) The crib protected the ears of corn from bears and other predators, and the ears were dried by the free passage of air between the unchinked logs (fig. 10.20). Log corncribs were the only buildings on many farms because the farmers needed no more, and elaborate Pennsylvania German forebay barns essentially disappeared south of Lexington.

Farmers needed larger farm buildings again when finally they pushed into the productive limestone valleys of southwestern Virginia and Upper East Tennessee, which became areas of intense experimentation. The rectangular corncrib was the basic module from which farmers developed a bewildering variety of structures by building larger cribs (fig. 10.21) and then by tacking all manner of side sheds and appurtenances onto them (fig. 10.22). Some farmers merely built a lean-to roof on one side of a small corncrib to shelter tools and other gear, but those on better land built larger cribs for storing hay, and they added lean-to sheds for cattle and equipment, first on one side, then on both.

Some farmers built stalls for their work animals on the ground floors of their enlarged cribs, and they stored hay in the lofts above them. A few drew on their familiarity with cantilever construction to extend their haylofts ten feet or more beyond the log cribs beneath them, to create grotesque frame overhangs on one, two, or even all four sides of the barn. Some farmers enclosed the space beneath these overhangs with outer shells of board siding, and you have to go inside the outer board cocoon to find the log core from which it has been expanded.

The taper and weight of logs limited the length and width of individual cribs to about thirty feet, but a farmer who needed more space simply built a second crib facing the first, with a hard-packed earthen driveway between them and a continuous roof above them (fig. 10.23). Either crib might be used for corn, for hay, or for work animals, or it might be divided to suit individual needs and whims, but the differences are too trivial to jus-

Fig. 10.21. A larger version of a log corncrib that was used to store hay.

Fig. 10.22.   A log hay barn that has been expanded by the tacking of sheds on either side. This barn is an ancestor of the livestock feeder barns of the Corn Belt (see fig. 10.24).

tify elaborate schemes of classification. Some people have confused these double-crib barns with three-bay barns, which they superficially resemble, but the driveway of a double-crib barn was not designed for threshing, and the threshing floor of a three-bay barn would not have solid log walls on either side.

The enlarged and augmented log cribs of southwestern Virginia and Upper East Tennessee became standardized into a new type of barn, a "livestock feeder barn," with lean-to board sheds flanking a rectangular log core with its principal opening in the end rather than in the side (fig. 10.24). This type of barn was not created full-blown by a single stroke of genius; it evolved through trial and error. The early models were cobbled together, but later barns of this type were built as single, unified structures, often with cores of boards rather than logs once the type had become standardized.

The area of origin in Virginia and Tennessee actually has few fully achieved livestock feeder barns, and its diversity of barns can be confusing because the region is littered with the results of other experiments that were not generally adopted. The successful type that became standardized and was disseminated to the Corn Belt was merely one of many in the area, but anyone familiar with Corn Belt livestock feeder barns will immediately recognize their local antecedents.

The core of this new livestock feeder barn was used to store hay, although its ground floor might have stalls for work animals. The side sheds were for other livestock, such as beef cattle, or for machinery (or almost anything else). The shift of the principal opening from the side, as it was in three-bay and Pennsylvania German forebay barns, to the gable end might seem trivial, but this shift signified a major alteration in the function of the barn that was associated with a fundamental change in the entire farming system. The barn was no longer a structure for threshing wheat, but a structure for storing hay, and the farming system was oriented toward the production and sale of livestock.

Corn Belt Barns

Settlers carried the idea of the livestock feeder barn northwest through Cumberland Gap, where Virginia and Tennessee meet Kentucky, to the rich limestone plains of the Bluegrass area of north central Kentucky, and on to the Miami River valley north of Cincinnati in southwestern Ohio, seedbed

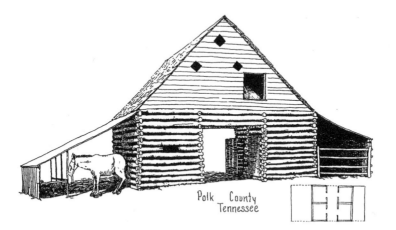

Fig. 10.23. A transverse-crib barn with lean-to side sheds. Despite a superficial resemblance, this barn should not be confused with a three-bay barn, because the central driveway was not used for threshing, and it has solid log walls on both sides. Reproduced by permission from Henry Glassie, "The Old Barns of Appalachia," *Mountain Life and Work* 40, no. 2 (1965): 29.

Fig. 10.24. The Corn Belt livestock feeder barn with a central core for hay and side sheds for livestock is a direct descendant of the single-crib log barns of Appalachia (see fig. 10.22).

of the Corn Belt. In the Miami River valley the livestock feeder barn from Appalachia came into competition with the Pennsylvania German forebay barn that settlers were bringing west with them.

Farmers in the Corn Belt adopted a modified form of the Pennsylvania German cropping system but rejected its forebay barns. Corn, grown in rotation with oats and clover, took the place of wheat as their most important crop. They stored the ears of corn in board cribs with slatted sides, reminiscent of the unchinked log cribs of Appalachia, and they used corn to fatten hogs and beef cattle. Pennsylvania German threshing barns were unnecessarily elaborate for the new Corn Belt farming system, however, and they lost out to the simpler livestock feeder barns from Appalachia (fig. 10.25). The Miami River valley still has many barns of both types, and farmers from southeastern Pennsylvania continued to build forebays on their barns wherever they settled, even though the forebay eventually was little more than a vestigial pent roof attached to the side or the end of a barn.

Livestock feeder barns with a core for hay and workstock and flanking sheds for other livestock and machinery were the norm in the Corn Belt until they were made obsolete by the changes that have revolutionized the region since 1960 (fig. 10.26). Most Corn Belt farmers have shifted from mixed crop-and-livestock farming to cash-crop farming based on corn and soybeans. They no longer have livestock, but they need to shelter their large and expensive new machines. The old wooden barns, which were built to last forever, can be converted to machine sheds only with great difficulty, and farmers have replaced them with shiny new multipurpose metal structures that can easily be converted from one use to another. Some romantics like to bewail the demise of fine old wooden barns, and they have even complained that modern metal structures are alien to the countryside, but no one can be a successful Corn Belt farmer today who is not able to appreciate the beauty of a multipurpose metal shed.

UNUSUAL BARNS

Regional variations in barn types reflect regional variations in farming, but each region also has considerable internal variation because of differences in the date of construction and size of farms, and the whims and foibles of individual farmers. During the twentieth century, barn types began to be homogenized by the institutional plans and designs that were

Fig. 10.25.   A perspective drawing of a Corn Belt livestock feeder barn, plan 72432, Midwest Plan Service, Iowa State University, Ames, Iowa. In his letter granting me permission to reproduce this drawing, John H. Pedersen, manager of the Midwest Plan Service, wrote, "I hope the illustration will not be billed as one for modern beef production. Although there are many buildings around essentially like the plan, it hardly fits modern production systems." *Sic transit.* . . . Copyright Midwest Plan Service, Ames, IA 50011-3080.

Fig. 10.26. A livestock feeder barn that does not fit the needs of modern production systems on the prairie of southwestern Minnesota.

pouring out of agricultural experiment stations, some for general-purpose farm structures, some tailored to the needs of specific farm activities. Some unusual types of barn transcend regional boundaries, some are widely distributed within given regions, and some are concentrated in subregions. Five of the better-chronicled unusual barn types are Dutch barns, Erie Shore barns, round barns, Barren County barns, and Cajun barns, but these are merely a sample.

Dutch Barns

The Dutch barns of the Hudson River valley and northeastern New Jersey are squat, square structures with steep roofs. They are wider than they are deep, and their roof ridges are more than twice the height of their side walls. Double doors in the gable ends lead to central threshing floors flanked by stalls for horses and cattle on side aisles, to which smaller doors in the corners of the gable ends give access. These barns had to be short to give an adequate draft for threshing, and they have an oddly truncated look. No other North American barns have threshing-floor entryways in their gable ends, and no others have livestock stalls and threshing floors on the same level under a single roof. No Dutch barns were built after the Revolutionary War, and the few that remain are little more than antiquarian curiosities; they have not influenced other American barns, although some people claim to see a superficial similarity between Dutch barns and the livestock feeder barns of the Corn Belt.

Erie Shore Barns

Peter Ennals first identified and named the Erie Shore barns in southern Ontario, but they have been built in widely scattered areas, mostly on small farms that need little barn space. One end of an Erie Shore barn has a through driveway where hay wagons could be sheltered while hay was forked into the loft above the stalls for cattle and workstock in the other end. Allen G. Noble has suggested that the design was developed around 1880 at the Michigan Agricultural Experiment Station as a cheap barn for a small farm, and it might well be one of the first successful barn types that was developed at an agricultural experiment station.

Round Barns

Round or polygonal barns definitely are one of the worst barn types ever advocated by an agricultural experiment station. They are so obvious and unusual that they excite curiosity, and every coffee-table book on barns seems to feel obligated to include a picture of the first round stone barn in the United States, which was built in Hancock, Massachusetts, in 1825. Few were built before 1905, however, when silos started to become important and the Illinois Agricultural Experiment Station began to champion the idea of round barns built around silos. Progressive farmers in many areas initially adopted the idea, but few round barns have been built since 1920, when their drawbacks had become painfully apparent.

The basic round barn has a silo in the center, livestock on the ground level, and a hayloft on the upper level that is reached by a barn bank. In theory it was easier to feed the cattle if they all faced inward toward the silo, but this theory, unfortunately, did not go quite far enough. It ignored what happens to the silage after the cattle have eaten and digested it, and cleaning out a round barn is a thoroughly unpleasant task, as any farmer who owns one will not hesitate to tell you in appropriately colorful language.

One might also wonder whether our rectangular society has an uneasy

Fig. 10.27.   A barn with slanted sides in Barren County, Kentucky.

sense of sacrilege about round barns because most of our round structures, such as capitols and cathedrals, are hallowed.

Barren County Barns

A distinctive and highly localized type of barn was built in the Kentucky Pennyroyal area in the 1920s, when the bottom dropped out of the dark-tobacco market and former tobacco farmers who began keeping small herds of dairy cows had to have barns. The county agent in Barren County was an intimidating personality. He "persuaded" many farmers to build barns whose ground floors had sides that rose vertically to a height of three to five feet, then slanted outward and upward at a thirty-degree angle for a few feet, and returned to the vertical five to eight feet above the ground (fig. 10.27). The slanted side formed the outer wall of a manger, or feed trough, into which hay could easily be pitched from the loft above. Barren County has many of these slant-sided barns, but they are few and far between outside the area where the persuasive county agent held sway, and most of those few were built by former residents of Barren County.

Cajun Barns

Malcolm Comeaux has identified a type of barn that was common on small, self-sufficient Cajun farms in south central Louisiana. Its core was a corncrib with a hayloft above and a small window in the rear through which corn could be shoveled into the crib. The entryway in the gable end was recessed six to eight feet, with side doors leading to flanking stall areas for three or four milk cows on one side and three or four work animals on the other. These barns strongly resemble the single-crib barns of Appalachia, but the recessed niche in front is unusual, and Comeaux argues that the idea for this type of barn was introduced to the area by early German settlers, from whom the Cajuns borrowed it.

THE LAST OF THE BARN

Today many barns are monuments to the past. They have outlived the functions they were originally built to perform: the hayloft is obsolescent, the threshing floor is obsolete, and who even knows how to hitch up a team of horses? Insofar as possible, the farmers who own old barns have modified them to meet their changing needs, but many were built so solidly that they are hard to modify. A few have been put to new and different uses, some have been bulldozed to get them off the tax list, and many have simply been allowed to go to quiet deaths.

Some old barns have been salvaged for their lumber. Suburbanites once were willing to pay outrageous prices for weather-beaten, worm-eaten boards from old barns to use for such things as picture frames and playroom paneling, but America's romance with barn boards seems to have ended. Each year an unknown number of old barns are guests of honor at barn fires, some accidental, some perhaps not. A barn fire is always good for a story in the local county paper; one correspondent covered all bases by noting that the neighbors "watched, were frightened by, or slept through the awesome midnight event."

11 *Other Farm Structures*

Fences and barns are ubiquitous on the agricultural landscape, but many other structures are also closely associated with agricultural activities. Some of these structures are related to livestock, some are connected with specific crops, some are in the fields, and some are at the farmsteads. Many of the structures for the collection, processing, and marketing of crops and livestock are not even on farms, but rather in the small towns that serve them.

Some farm buildings are multipurpose structures that are put to a variety of uses at different times of the year, but some, such as oast-houses and mint stills, serve only a single, specific function. The appearance of some structures, such as tobacco barns, is determined by their function, but farmers have far greater leeway in designing others, whose appearance may vary from region to region and from one culture group to another within a region. Some of the more mundane structures, such as garages and privies, do not seem to have challenged the decorative ingenuity of farm families; these structures have much the same nondescript appearance in all parts of the country and among all culture groups.

Some of the newest, most standardized, and least distinctive structures on modern farms are machine sheds. A modern multipurpose machine shed is a highly efficient and functional metal box. At least one side consists mainly of sliding doors to facilitate the movement of machinery in and out; a well-equipped workshop with a small window may be tucked in one corner. They are not identical, even though they look as though they might all have been bought from the same mail-order catalog, and many city folk feel that "when you've seen one, you've seen 'em all," whether you are in Washington, Florida, Maine, California, or anywhere in between. Their differences seem to be mainly matters of size, color, trim, and the placement of doors.

STRUCTURES FOR LIVESTOCK

The farmer who has livestock must protect them from wind, snow, cold, rain, excessive sunshine, and extremes of temperature. The structure that houses the animals should permit the farmer to feed and handle them with convenience and safety, both to people and to beasts. It should have an adequate water supply, and it should permit easy cleaning and removal of manure. On many small farms in the old days, horses, cattle, hogs, and other livestock were all kept in a single general-purpose barn, which was also used to store hay and grain. Larger farms might have separate horse barns for the workstock, but the replacement of horses by tractors has made the old horse barn redundant, just as increasing specialization has made general-purpose barns obsolete, and both types have been converted to other uses, torn down, or allowed to fall down.

Feedlots

No feed produces finer beef or pork than corn, and American farmers have been "walking their corn to market" by using it to fatten cattle and hogs ever since the government made them stop marketing it in liquid form. Four of every five ears of corn grown in the Corn Belt once were used to fatten hogs or cattle to market weight, a process known as feeding or finishing. Most of the cattle fed by Corn Belt farmers were raised on ranches in the grassland areas of the West, which could produce lean "4-H cattle—all hooves, horns, hide, and hair," but not the kind of tender meat that American consumers demanded, so ranchers had to sell their lean "feeder cattle" to farmers in the Corn Belt to be finished.

Farms in the western Corn Belt had stoutly fenced feedlots that could hold several hundred feeder cattle (see fig. 9.14). The farmers bought cattle after harvesting their crops in the fall, fed the cattle to slaughter weight during the winter, and shipped them off to the stockyards in the spring before they had to begin crop work again. Ideally the cattle were placed on feed at a weight of around 650 pounds, fed for 120 to 150 days, and sold fat at a weight of 1,050 pounds. Most farm feedlots had loading ramps and chutes, or squeeze gates, where animals could be held fast for dehorning, hoof trimming, and other veterinary work, and many had scales where the cattle could be weighed when they arrived and departed.

Since World War II, cattle feeding has spread to irrigated areas in the West, where it has been stimulated by the growth of the population, an increased demand for meat, and an abundance of feed. It has become big business, with huge new specialized feedlots holding tens of thousands of cattle, in sharp contrast to the smaller farm feedlots of the Corn Belt (fig. 11.1). The new feedlots are essentially factories that buy cattle and feed, the raw materials they need to produce beef, and they grow no feed of their own. They are called dry lots because they have no grass or other vegetation, but they have water and feed troughs that are filled mechanically by self-unloading trucks driving slowly past.

Corn, grain sorghum, and barley are the principal grains fed in western feedlots, and alfalfa hay, corn silage, and other hay provide most of the necessary roughage, but the cost of feed is such a major factor in the success of feedlots that their operators will feed almost anything they can buy cheaply and the cattle are willing to eat. Cottonseed hulls and cake are staple feeds in southern irrigated areas, as are sugar beet tops and pulp in northern areas. The list of feeds also includes spent mash from distilleries, citrus pulp from juice-concentrating plants, grape pumice from wineries, almond hulls, peanut meal, olive meal, pear pulp, waste from vegetable- and fruit-canning plants, and surplus raisins, prunes, carrots, and cantaloupes.

During the 1960s a major new cattle-feeding area developed on the High Plains of the Texas panhandle and in southwestern Kansas. A reduction in the cost of shipping cold meat to market enabled the meat-packing industry to get out of major metropolitan areas, and the increased costs of shipping feed grains and cattle encouraged it to construct new packing plants near the sources of supply. Eight new plants increased the capacity of the High Plains from 0.4 million head in 1960 to 2.6 million head in 1969, and the number of cattle on feed in the area increased from 100,000 to 950,000 because of the retention of cattle that formerly had been shipped to feedlots in other areas. Abundant feed was available for them in the form of grain sorghum, which local farmers had started to grow on a large scale in the late 1950s because it is well suited to the semiarid environment.

Dry lots are not restricted to beef cattle. Dry-lot dairy farming is an efficient strategy for producing milk near large cities in the West, where the competition for land is intense. In 1994 an area of thirty square miles in the Chino Valley east of Los Angeles had 260 dairy farms with an average of 925 cows each, when the average family-sized dairy herd in Wisconsin was only 50 cows. Rectangular bales of alfalfa hay trucked in from distant irrigated areas are stacked high beside the dairy dry lots (fig. 11.2). Twice a day the cows trudge to a milking parlor, a small building with a pit down

Fig. 11.1. A beef-cattle feedlot on the Great Plains.

the center and stanchions and milking machines on either side. The worker in the pit secures the heads of the cows in the stanchions, places their rations of concentrated feed in front of them, and attaches the milking machines. After they have been milked, the cows return to the dry lot, where they eat alfalfa hay and produce more than a billion pounds of manure each year.

The people who live near them complain constantly about the odors, flies, and noises of dry-lot dairy farms, which have difficulty disposing of the manure that the cows produce in such prodigious quantities unendingly and at such a rapid rate in such a small area. The disposal problem of the dry-lot dairy farms in the Chino Valley is equal to the sewage-disposal problem of the City of Los Angeles.

Poultry Houses

Once, farmers kept a few chickens and hogs in various and sundry nondescript sheds around the farmstead. They could convert almost any old building into a poultry or hog house without too much trouble if they needed more space. In recent years, however, animal breeders have trans-

Fig. 11.2.   Dairy cows on a dry-lot dairy farm in California, with bales of alfalfa hay under the shed in the distance.

formed poultry and hogs into efficient but sensitive meat-making machines, and farmers must shelter them in special climate-controlled structures. Even hogs, despite their appearance and behavior, are surprisingly delicate creatures, and they need tender loving care. The new poultry and hog houses, even more than dry lots, are factories that use the animals as machines to convert raw materials—corn and soybeans—into a marketable product—meat—although it is unusual for factories to use machines that are designed to be eaten after they have done their job. Turkeys and eggs also are produced in similar purpose-built structures; modern egg farms are truly enormous, with four million hens or more each laying three eggs every four days onto conveyor belts.

Modern poultry houses are one-story metal structures, hundreds of feet long and up to fifty feet wide (fig. 11.3). At one end stand round metal feed bins with conical bottoms that taper down to load automatic feeding systems. The upper parts of the side walls of the houses are wire-mesh screens with canvas curtains that can be raised or lowered for temperature control. The temperature must be kept just right because the birds convert feed into body heat rather than meat if they are too cold, and on hot, muggy days they seem to melt. A water fogger in the roof puts out a cooling spray in summer, and brooder stoves heat the house in winter.

The poultry business has become vertically integrated. Poultry farmers are merely one link in the chain connecting the hatchery that produces baby chicks, the feed company that buys the grain and prepares the feed, the processing plant that dresses the birds for market, and the retail outlets where they are sold. Before World War II, individual farmers bought their own chicks and feed, but the price fluctuated so rapidly, and their profit margins were so low, that they were reluctant to risk their own capital in the poultry business.

The vertical integrators, usually feed companies or processors, stepped into the picture with the risk capital necessary to forge complete chains linking hatcheries, grain elevators, feed mills, farmers, processing plants, and market outlets. The integrator supplies baby chicks, feed, management, processing, and marketing, and the farmer provides the land, labor, buildings, and equipment. Poultry farmers have contracts that guarantee them a fixed

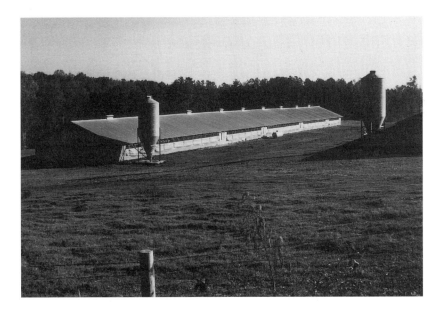

Fig. 11.3. A broiler house for raising chickens.

Fig. 11.4. Young chicks in a broiler house.

Fig. 11.5. Seven weeks later the birds from fig. 11.4 have grown so large that one can hardly walk through the house.

price for the birds they fatten. They like this guaranteed income for their labor, but they resent that the company retains the right to make all management decisions, and that it can give them orders on their own farms.

Poultry houses have steadily gotten larger. In 1940 a fifteen-hundred-bird house was considered large, but ten-thousand-bird houses became standard in the 1950s, and by 1990 feed companies would not even deal with farmers whose houses held fewer than twenty thousand birds. The rule of thumb is that a bird needs about one square foot, so a twenty-thousand-bird house might measure forty by five hundred feet. Farmers receive day-old baby chicks from the feed company (fig. 11.4), and fatten them to a market weight of four and half pounds in seven weeks (fig. 11.5). The farmers leave the lights on inside the houses all night long to encourage the birds to keep eating constantly. The interior of a poultry house is a piano-wire jungle because the feed troughs, watering pans, and stoves are suspended from the roof so they can be raised by hand winch out of the way when the catchers come to collect the fattened birds.

Hog Houses

Hog production was the last major form of livestock production to develop large-scale industrial operations in purpose-built structures, but when it finally did so it did so with a vengeance. In 1985 an operation with five hundred sows was considered huge, but ten years later thousand-sow operations had become the norm.

Hog houses, like poultry houses, are long, low, narrow metal structures, but unlike poultry houses their sides usually have large circular openings a few feet above the ground for ventilator fans because hogs generate tremendous amounts of body heat (fig. 11.6). A sow prefers a temperature around 65°F, and she starts to feel stressed when it gets above 80°F. Hog farmers usually have several buildings that are divided into separate pens, and they move batches of animals from building to building as they grow (fig. 11.7). Near the building complex is a large settling lagoon into which manure is pumped from the houses (fig. 11.8). The lagoon is lined with solidly packed clay or sheets of heavy plastic to prevent seepage and groundwater contamination.

The sows give birth in the farrowing house. They are poor mothers, and the baby piglets must have protected areas where their mothers will not roll over and crush them. The farmers move the piglets to the weaning house when they are old enough, and then to the finishing house for final

Fig. 11.6.   Ventilator fans on hog houses.

fattening. Hogs eat like hogs, and they attain their sales weight of two hundred twenty-five to two hundred sixty pounds by the time they are six months old.

A hog operation should be at least a mile from any other hogs as a precaution against disease, because hogs are highly susceptible to a variety of deadly ailments. Most producers are understandably nervous about the germs a visitor might bring, and they attempt to control disease with a strict "shower in, shower out" policy. Every single person entering or leaving any hog building must strip to the skin, take a thorough shower, wash his or her hair, and put on a set of completely clean clothes.

Hog producers feel beleaguered and wary of visitors not only because of the threat of disease but also because of complaints about pollution. Hogs produce four times as much solid waste as people do, so a thousand-sow operation with ten thousand feeder pigs will produce as much solid waste as a city of forty thousand people. This waste is pumped into open-air lagoons where anaerobic bacteria will break down organic matter. Critics believe that seepage from these lagoons can pollute groundwater, and no one can deny that they pollute the air, because they stink to high heaven.

Hog producers have also been criticized by animal-rights activists, who have done their best to stir up public outrage against confined production of poultry and hogs. The producers respond that it is in their own best interests to keep their animals as healthy and as happy as possible because they are not going to make money from animals that are sick or so unhappy that they refuse to eat.

Market demands are pushing the pork business toward fewer but larger farms. Consumers want leaner meat in more-convenient packages, and livestock breeders have developed animals that grow faster and need less feed to produce leaner hogs. Traditional small-scale hog farmers have been reluctant to invest in new production technologies, however, and pork processors have contracted with large-scale farmers or even started their own pro-

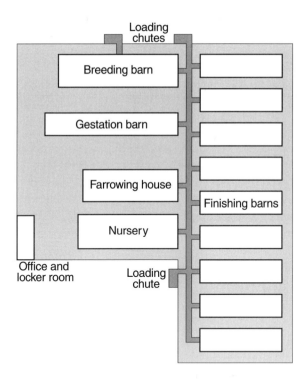

Fig. 11.7. The layout of a thousand-sow farrow-to-finish hog operation. The sows spend five to six weeks in the breeding barn, 109 days in the gestation barn, and four weeks in the farrowing house before they return to the breeding barn. The piglets are moved from the farrowing house to the nursery, where they stay for six weeks until they weigh 50 pounds. Then they are moved to one of the nine finishing barns, where they eat like hogs for sixteen weeks until they weigh 260 pounds and are ready to be processed.

Fig. 11.8.   An aerial view of a thousand-sow farrow-to-finish hog operation, with a manure lagoon just beyond and to the left. Reproduced by permission of Marty Manuel of Carroll's Foods of Virginia, Inc.

duction operations to ensure a steady supply of hogs of the size, shape, and quality they need.

A few large contracts are easier to manage than many small ones, and large hog farms enjoy significant economies of scale; in 1990 the cost of producing one hundred pounds of pork was sixty dollars in 150-head operations and forty-five dollars in 10,000-head operations, although large operations also entail much greater risk and require far greater management skills. Small-scale hog farmers in some traditional pork-producing states have tried to block technological innovation by enacting laws restricting large operations, and the pork industry has been forced to seek new areas with less restrictive laws.

North Carolina has become the nation's second-most-important hog-producing state, following only Iowa, and the erstwhile Dust Bowl of southwestern Kansas and the Oklahoma and Texas panhandles has an astonishing concentration of hog farms (fig. 11.9). Seaboard Farms, Inc., has invested three hundred million dollars in developing more than one hundred hog farms in the area, and in a meat-packing plant in Guymon, Oklahoma, that will process four million hogs a year, one-seventh as many as the entire state of Iowa. Other companies that have been attracted to the area include Texas Farms, Inc., a subsidiary of Nippon Meat Packers, the world's fourth-largest meat-packing company. Texas Farms has invested a hundred million dollars in a farrow-to-finish hog operation near Perryton, Texas, to produce 540,000 hogs a year to be processed locally for export to Japan. The company has built 350 barns on four separate sites (fig. 11.10). Clusters of nine to twelve finishing barns house 900 hogs each, with smaller clusters containing farrowing houses and nursery barns (fig. 11.11).

The buildings are flushed weekly to remove the manure that has fallen through their slatted floors. The wastewater goes to a primary lagoon. Some of the wastewater is recycled to flush the houses after the solids have settled out of it, and the rest goes to a secondary lagoon to evaporate. The company was attracted to the panhandle area by its abundant supplies of feed grain, dry climate, steady winds, and high evaporation rates. Its relatively

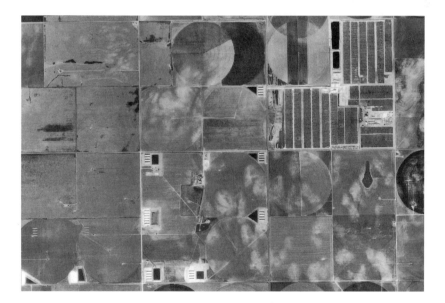

Fig. 11.9. An excerpt from an aerial photograph (U.S. Department of Agriculture NAPP 8464-165, 2-10-95) of a beef-cattle feedlot, hog farms, and center-pivot irrigation circles twenty miles north and a bit west of Guymon, Oklahoma. The long, narrow strips of the feedlot occupy three-quarters of a section (square mile) in the upper right corner. The section catercornered from the feedlot has six hog farms with four to six parallel buildings, and four more hog farms were being developed in the section just to the west. The area of darker tone next to each hog farm is a manure lagoon. Triangular lagoons occupy the corners that are not irrigated by center-pivot systems.

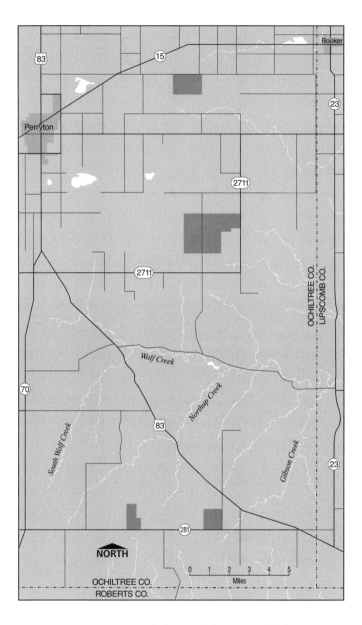

Fig. 11.10. In 1995, Texas Farms, Inc., had acquired four tracts of land near Perryton, Texas. It planned to put five or six hog operations on each square mile, and expected to produce 540,000 hogs a year to be processed locally for export to Japan.

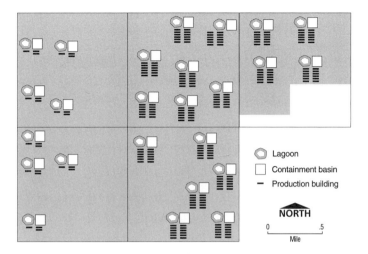

Fig. 11.11.   Hog operations on the largest tract of land acquired by Texas Farms, Inc., near Perryton, Texas (see fig. 11.10). The smaller clusters are for breeding, and the larger clusters are for finishing. Each cluster has a pair of lagoons for evaporating the manure.

sparse population is a plus because it has fewer people to complain about odors and environmental contamination.

STRUCTURES FOR CORN

Some farmers, especially those who are hard-pressed for cash, will haul their corn directly from the harvest field to the local grain elevator and sell it. If all farmers sold their corn at harvest time, however, they would flood the market and depress the price when it is already at its lowest, so most farmers have seen fit to invest in facilities for storing at least part of their crop until they think the price is right. Furthermore, much of the corn crop is grown to be fed to livestock on the farm, not to be sold, and farmers who feed livestock must have storage both for the corn they produce and for any they may have to buy to supplement their own production if they fall short.

Corncribs

Corn is the quintessential American crop. The pioneers fed the grain to their animals, they ground it up to make cornmeal, they leached it in lye

Fig. 11.12.   Corn in shocks on an Appalachian hillside.

to make hominy, and they boiled it with beans to make a tasty dish of succotash. The Native Americans showed them how to grow everything they
needed—corn, beans, and cucurbits—on the same small patch of cleared
land. The tall stalks of corn supported the bean vines, and pumpkins,
squash, gourds, and cucumbers grew on the ground beneath. In autumn,
when the ears were ripe, the pioneers cut the stalks of corn and stacked
them in shocks in the field (fig. 11.12). During the winter they broke the
ears from the stalks and shucked off the husks from the ears.

Corn can be stored almost indefinitely when its moisture content is 15
percent or lower, but it must be protected from birds, squirrels, rats, rac-
coons, and other pests. At first the pioneers stored their corn in the attics of
their houses, but by 1650 they were already building simple cribs of unchinked logs (see fig. 10.20). Air movement through the cracks between the
logs dried the corn that was stored in the crib. Some cribs had straight sides,
but some had V-shaped sides that flared upward and outward. From the
end these coffin-shaped corncribs looked top-heavy, but they shed rain better than cribs with straight sides. Wooden boards or laths replaced logs
when cheap lumber became available, but they still were spaced far enough
apart to allow air movement to dry the corn (fig. 11.13).

Farmers in the Corn Belt husked their corn by hand in the field until mechanical corn pickers began to replace hand harvesting in the 1920s. The farmer tore the ears of corn from the stalks, husked them, and threw them into the trailing wagon that hauled them back to the crib. The early cribs were filled by hand, and they could be no higher than the height at which a person standing on a wagon could conveniently throw ears of corn with a wide scoop shovel. Farmers who needed more space either made their cribs longer or built two cribs facing each other. The standard double crib of the Corn Belt evolved when farmers extended the lean-to roofs of a pair of facing cribs into a single gable roof that sheltered the passage between them.

The development of mechanical grain elevators in the 1930s enabled farmers to build taller corncribs. A grain elevator is a trough containing a continuous belt that is driven by the power takeoff of a tractor parked beside it. Attached to the belt is a series of cups or blades that catch the ears of corn and lift them to a cupola on top of the crib, from which chutes direct them to the desired compartment. The standard double crib in the Corn Belt was thirty-two to forty feet long, with slatted cribs eight to ten feet wide on either side of a central passage (fig. 11.14). The area above the

Fig. 11.13. A corncrib with sides of wooden laths spaced far enough apart to allow air movement to dry the corn. The building material is different, but this corncrib is functionally similar to the log corncribs of Appalachia (see fig. 10.20).

Fig. 11.14.   A standard double corncrib of the Corn Belt.

passage had bins with solid sides that could hold loose grain. Some farmers built cribs of hollow clay tile or concrete blocks, and in a pinch they even made temporary cribs of wire mesh or snow fence, but slatted wood remained the norm.

Grain Bins

Traditional wooden corncribs with slatted sides are rapidly vanishing from the rural landscape because self-propelled combines with prong-type corn heads, which became common in the 1960s, made them obsolete almost overnight. The combine tears the ears from the stalks, shells the kernels of grain from the cobs, and stores the grain in a bin behind its cab. When the bin is full the farmer unloads it through an ungainly-looking spout on its side into the vehicle that hauls the grain back to the farmstead. Loose grain would pour through the slatted sides of traditional corncribs, and farmers have had to replace them with cylindrical grain bins made of corrugated metal (fig. 11.15). The farmers fill their bins with augers, long metal tubes containing screws that turn to lift the grain.

Moist grain would mold in an enclosed metal bin, and farmers must dry their grain before they can store it. The grain dryer is at the center of a nest of bins. Towering above it is a "leg" containing a cup elevator that lifts the dried grain to a distributor head at the top, where it is directed into a long metal tube that will carry it to the storage bin. Each bin can be unloaded at the bottom by an auger that carries loose grain back to the base of the elevator leg. The new grain-harvesting and grain-handling systems can manage soybeans, the other major crop on cash-grain farms in the Corn Belt, as easily as they handle shelled corn (fig. 11.16).

Stills

In some of the less accessible parts of the United States it used to be customary to process part of the corn crop on the farm in order to reduce its bulk and convert it into a product of greater value per unit of weight before transporting it to market. This practice was especially well developed in Appalachia, where it was based on the technical competence that the early Ulster Scot settlers brought with them from Ireland. Although it was eminently reasonable from the point of view of economics, this practice found disfavor with the legal authorities, who saw fit to tax the resulting product at a rate that seemed quite inequitable to the producers. Such fric-

Fig. 11.15. Traditional wooden corncribs have been superseded by grain bins of corrugated metal.

Fig. 11.16. Cash-grain farmers need a battery of grain bins, with a central elevator leg that lifts loose grain to the distributor head at the top, whence it is fed through tubes to the appropriate bin.

tion resulted that it was difficult, and sometimes downright dangerous, to photograph and describe the structures associated with this particular practice. Hovering helicopters and infrared heat-sensing devices have essentially consigned stills to the dustbin of history.

Silos

Silos became common structures for storing corn in the northern tier of states that shifted to dairy farming after the wheat frontier had passed through on its way westward. The growing season in these areas is so short and so cool that the corn crop often was damaged by frost before the grain had time to ripen, but the stalks and leaves could be fed to dairy cows. Dairy farmers cut the green stalks, stacked them in shocks in the field to dry, and fed bundles of dry corn to their cows through the winter. Cows fed dry fodder usually would "go dry," or stop giving milk, in November, and not start again until the first flush of new growth in April. They produced the most milk in the spring and early summer, but people drank the most milk in the fall and winter, when production was lowest and prices were best for farmers.

Agricultural scientists encouraged farmers to switch from dry-fodder corn to silage for winter feed, because succulent silage kept the cows producing milk through the winter. The farmers cut off the entire corn plant close to the ground before the grain was fully mature, chopped it into small pieces, and stored them in enclosed silos. The first silos were made of wood, but cement soon replaced wood, and towering cylindrical cement silos became the hallmarks of the dairy landscape. By World War I nearly every dairy farm had at least one, and many farms had several.

Silage yields have increased so dramatically in recent years that farmers need to plant only about half as much land in corn to fill their silos as when the silos were built. Some have built additional silos to hold their increased yields, but others have elected to plant new short-season varieties of corn they can harvest for grain on the land they no longer need in order to fill their silos. Their corn acreage is too small to warrant investing in the new technology for harvesting and handling grain, however, and some dairy farmers still use the kinds of mechanical corn pickers and wooden corncribs that have become obsolete in the Corn Belt.

Corn is the principal silage crop in the United States, but farmers also make silage from grass, sugar beet tops, waste from fruit- and vegetable-processing factories, and other plant matter. Not all silos are vertical. Some farmers store silage in pits or trenches, or in aboveground bunkers constructed of concrete or treated lumber. They cover these horizontal silos with sheets of black plastic and weight the plastic down with old automobile tires to keep it from blowing away. Horizontal silos are less efficient than vertical silos because more of the silage spoils, but they are cheaper and easier to build, and cattle can eat directly from the silos instead of having to be fed.

Big, blue, glass-lined metal silos came on the market after World War II (fig. 11.17). Glass bonded to the inside of the metal forms a gastight container, and it preserves silage just as glass jars preserve food. These silos unload automatically from the bottom, and the farmer can deliver silage directly to feed troughs at the touch of a button. The silage comes out of the silo in almost the condition in which it was harvested. Some farmers use these silos for high-moisture corn, which they harvest when the moisture of the kernels gets down to 35 percent. High-moisture corn has the same feeding value as dry shelled corn, and farmers save the expense of having to dry it, but it spoils so quickly that they must use it immediately after they take it out of the silo.

Many farmers have become increasingly reluctant to trust animals to harvest their forage crops; they prefer to do the job themselves because they can do it more efficiently. The grazing animals waste some of the crop by

Fig. 11.17.   Glass-lined metal silos are gastight, and they preserve fodder just as glass jars preserve food. The company that makes them also makes squat blue metal tanks for holding manure.

trampling it, and soil a surprising amount of it with their droppings. They have to graze it over a period of time, consuming some when it is still green and immature and some when it is overripe and past its prime. The farmer can harvest the entire crop at just the right time, store it in a silo, and dole out the proper amounts in feedlot troughs. Many farmers, even in such traditional grazing areas as the dairy country of Minnesota and Wisconsin, have begun to switch to feedlot feeding year-round, and now use forage harvesters rather than cattle to harvest their crops. The animals never set foot on pasture, and farmers have begun to remove their permanent fences, which are no longer necessary; if they need to enclose a piece of ground temporarily, the farmers can "slap a hot wire [electric fence]" around it.

The new metal silos are more expensive than conventional cement silos, and some farmers dislike catching the eye of the tax assessor so easily. The company that makes them also makes squat blue metal tanks for holding liquid manure until it is time to spread it on the fields (see fig. 11.17). These tanks must have secure fences and guards because more than once, curious small children have fallen into them and drowned—surely one of the most horrible deaths imaginable.

STRUCTURES FOR TOBACCO

The impact of tobacco on farm income and on the rural landscape is quite disproportionate to its acreage. Only nursery and greenhouse crops are more valuable than tobacco, which returns more than twice as much per acre as vegetables do, and more than fifteen times as much as corn or soybeans. No cultivated crop has smaller seeds. They are smaller than the period at the end of this sentence, and a single teaspoon can easily hold more than a hundred thousand seeds. They must be planted in specially prepared seedbeds that are covered with sheets of cheesecloth or plastic to protect the delicate young plants. Each acre of tobacco needs a seedbed one hundred feet long and nine feet wide, and the long, narrow seedbeds are distinctive features of the tobacco landscape in the spring (fig. 11.18). Transplanting the seedlings to the fields is only the first of many labor-intensive activities necessary to produce tobacco, and farmers are already busy preparing the seedbeds for next year before they have sold this year's crop.

Each tobacco-producing area yields a unique type of tobacco. Each type

Fig. 11.18. Tobacco seedbeds are covered with sheets of cheesecloth to protect the young plants. Ideally a seedbed is on ground that has never grown tobacco, to minimize damage from diseases and insects.

Fig. 11.19. The interior of a tobacco barn has a lattice of tier poles on which the loaded tobacco sticks are hung.

has a particular use, and it cannot be used for anything else. The United States Department of Agriculture officially recognizes no fewer than twenty-six different types, but most are minor or of only historic interest. The leaves of each type must be dried, or cured, before they are ready for sale, and each tobacco district has its own distinctive curing barns. Before they are placed in the curing barn the leaves are attached to tobacco sticks, an inch square and four and a half feet long. The interiors of tobacco barns are more alike than their exteriors because most of them have an interior lattice of cross-pieces called "tier poles" on which the loaded tobacco sticks are placed (fig. 11.19).

Bright Tobacco

Farmers in the Carolinas and Georgia used to cure their tobacco by baking it in small, cube-shaped barns measuring sixteen to twenty feet on a side (fig. 11.20). Once a week during the fall harvest season the farmers picked the one or two leaves that had ripened on each plant, tied them to tobacco sticks, and hung them in the barn. One side of the barn had a squat furnace, usually under a shed roof, from which twelve-inch flues of

Fig. 11.20.   Farmers in the Carolinas and Georgia once flue- cured their tobacco by baking it in small cube-shaped barns.

sheet metal extended across the barn floor. The farmer fired up the furnace when the barn was full, and the heat from the flues baked the leaves for three or four days to give them the bright golden-yellow color that gives the leaf its name. Then the sticks had to be removed from the barn to empty it for the next week's picking, and the cured leaves had to be untied from the sticks before they could be taken to market.

In the 1960s farmers began to mechanize the production of Bright tobacco. They developed mechanical harvesters that could be set to cut off all leaves at a given height above the ground, which presumably had ripened at about the same time. They developed new metal bulk barns that use forced air rather than convection to bake the leaves (fig. 11.21). In bulk barns, which look like semitrailers, they could clamp the harvested leaves in metal racks instead of having to tie them to sticks, and they could pack them into the barn much more tightly. Bulk barns have completely replaced the old wooden flue-cured-tobacco barns. It is still called flue-cured tobacco, even though precious little is flue cured anymore, and in the tobacco trade it is still called Virginia tobacco, after the state in which it was first produced, even though precious little is grown in Virginia anymore.

Fig. 11.21. Metal bulk barns that use forced air have superseded traditional flue-cured-tobacco barns, which depended on convection.

Fig. 11.22. Air-cured-tobacco barns have vertical doors that can be opened or closed to control the curing process.

Fire-Cured Tobacco

In western Kentucky and Tennessee some farmers used to cure their tobacco with a mixture of heat and smoke. After they had filled their barns with tobacco they placed small piles of hardwood on the earthen floors and set fire to them. When the fires were burning well the farmer smothered them with damp sawdust to create a dense pall of smoke that imparted a distinctive flavor to the leaves. These barns should have been airtight, but few of them were, and the smoke billowing out through their cracks was a spectacular sight. Farmers liked to site them in or near woods to reduce the loss of heat and smoke. Today fire-cured tobacco is little more than a historical curiosity, but it was an important ingredient of snuff and chewing tobacco in the days when some people could tolerate stronger tobacco.

Air-Cured Tobacco

The other types of tobacco produced in the United States are cured without heat in rectangular wooden barns with numerous ventilator doors (fig. 11.22). Every third board on the sides of these barns is on hinges. These hinged boards are doors that can be opened or closed to control the humidity and the temperature inside the barn. The best site for an air-cured-tobacco barn is the crest of a ridge or some other exposed location that gets the full drying benefit of wind from any direction. For most of the year these barns are merely simple rectangular structures, but they are quite striking at the end of summer when their ventilator doors are open.

Air-cured-tobacco barns vary greatly from area to area. Most have vertical doors, but some doors are horizontal. Some are not painted at all, but they are painted red in Wisconsin, white in Pennsylvania, and two-tone in the Bluegrass area of Kentucky, where the ventilator boards are painted a different color from the rest of the barn. Their size is proportional to the tobacco acreage of the farm, and farmers with only small tobacco acreages will hang their crop in any space available.

Attempts to mechanize the production of air-cured tobacco have had only limited success, and the crop still is grown and harvested by traditional hand methods. At harvesttime, workers use a hatchet to cut off each plant near the ground, and they impale it on a tobacco stick. They slip a metal spearpoint onto the beveled end of the stick to make it easier to impale the tough stalk of the plant. After the leaves have been cured they must be stripped from the stalks, and most air-cured-tobacco barns have

Fig. 11.23. Entire tobacco fields that produced cigar-wrapper tobacco were covered by cheesecloth shades to simulate a moist tropical climate.

Fig. 11.24. "Little hay men" standing in a field in the Austrian Alps.

small stripping sheds, either beside the barns or in the basements, where workers can tear the leaves from the stalks and gather them into bundles to be taken to market.

Shade-Grown Tobacco

A special type of tobacco that was used only for making the eye-catching outer wrappers of cigars once was produced under enormous cheesecloth shades that covered entire tobacco fields in the Connecticut River valley of Massachusetts and Connecticut and in Gadsden County, Florida (fig. 11.23). The picturesque shades protected the plants from direct sunlight and simulated a moist tropical climate by increasing humidity and reducing evaporation and wind velocity. In the spring, workers attached the shades to wires stretched between posts nine feet tall, and they took the shades down after the crop had been harvested in the fall. The sides of the shades could be raised to allow workers and equipment into the fields. Shade-grown-tobacco farms were large because they required a substantial capital investment, and their air-cured-tobacco barns were proportionately large. In Florida the farmers used their barns during the off-season to house and feed the cattle that produced the manure they needed to fertilize their sandy soil.

HAYSTACKS AND HAY BALES

Grasses and legumes are grown more widely, and on a larger acreage, than any other crop. The field on which they are grown may be used as a pasture if it is suitably fenced, and the crop may be harvested by grazing livestock, but much of it is stored as hay or silage for winter feed. In the United States swaths of hay, which may be turned occasionally, are left on the ground to dry, but in the moister parts of Europe farmers have learned from sad experience that they cannot count on the sun to cure the hay crop unless they keep it off the damp ground, and they have developed a variety of ingenious devices for doing so. In Alpine areas farmers lop off the branches of young trees fairly close to the trunk, drive the trunk into the ground, and hang hay on the stubs of the branches. From a distance these "little hay men" look almost like people standing in the fields, whence their colloquial name (fig. 11.24).

Many farmers in Europe stack their hay on wooden tripods to keep it off the ground. Farmers in parts of Denmark prefer quadripods, which per-

Fig. 11.25.   Hay drying on quadripods in Denmark.

Fig. 11.26.   Hay being dried on a "clothesline" at the head of a fjord on the moist west coast of Norway.

mit air to flow beneath the drying plants as well as around them (fig. 11.25). Quadripods are especially useful for drying the tops of sugar beets. Farther north, in the cooler and damper areas of Norway and Sweden, farmers hang hay on ropes strung between posts they have driven into the ground, as if on a clothesline (fig. 11.26), or on temporary wooden racks they construct in their fields (fig. 11.27). In the eastern Alps farmers hang hay on horizontal wood crosspieces, spaced about a foot apart, whose ends are attached to sturdy upright poles that also support a sheltering roof over the drying crop (fig. 11.28). In earlier times some farmers built these hay-drying racks against the sides of their barns; later they enclosed them, and these enclosures became the ancestors of the forebays on Pennsylvania German forebay barns.

After it has been cured, hay may be stored loose or in bales. Ranchers in some parts of the West have haystacking machines that can pile up a field stack containing several tons of hay in a short time. California has mainly square, flat-topped stacks; high, round-topped stacks are more common in the intermountain states, and the Great Plains have low, round-topped stacks. In Appalachia farmers used to build field stacks of hay around a central pole for stability, but in the Midwest farmers preferred to store loose hay in the lofts of their barns.

Rectangular bales of hay are more weather-resistant than loose hay, and farmers may stack them in the fields or store them in barn lofts. Large-scale hay producers in the Imperial Valley of southern California have special machines that pick up hay bales, load them onto a trailing platform, haul them to the roadside, and build them into enormous stacks that wait for the trucks that will haul them to dry-lot dairy farms or five-acre horse ranchettes around Los Angeles (fig. 11.29). In the Midwest many farmers now use machines that ball hay into huge circular bales that weigh half a ton or more. Some farmers leave these bales sitting picturesquely in the fields; others haul them back to the farmstead and line them up neatly near the lot where they will be used.

OTHER STRUCTURES IN THE FIELDS

Orchards and Vineyards

Most fruit and nut crops are grown on trees or bushes that are more permanent than field crops, which must be replanted each year. The trees and bushes are evenly spaced in straight rows for ease of management, and an orchard or vineyard, once established, remains a distinctive feature of

Fig. 11.27. Temporary wooden racks for drying hay in central Sweden.

the landscape for many years. Producers of delicate fruits like to reduce the risk of frost by putting their orchards on sloping land, for good air drainage, and near temperature-modifying bodies of water if at all possible (fig. 11.30).

Grove owners prefer smaller and shorter plants because they are easier to pick, but each variety has its own requirements, and the size and spacing of individual plants varies from crop to crop. A mere list of the major fruit (apples, peaches, plums, pears, cherries, oranges, grapefruit, lemons, avocados, figs, dates) and nut (pecans, almonds, walnuts) crops hints at the variety and complexity of orchard layouts, whose proper treatment would require an entire treatise.

Vineyards are a special case because the vines are trained to grow on trellises of wire strung between posts set in the ground (fig. 11.31). The shading of the interior leaves strongly influences the quality of wine grapes. In the traditional vineyard areas of Europe the vines are planted in low, narrow, closely spaced rows that restrict the growth of leaves. Few interior leaves are shaded, and the grapes ripen to their fullest. In newer vineyard areas in other parts of the world the rows are spaced farther apart to permit the use of machines, and the vines would be too leafy to produce top-quality wine grapes if the growers had not developed a different system of trellising.

Hopyards and Oast-houses

The hop plant is a vine whose conelike buds, or catkins, are dried and used to give beer its bitter taste and aroma. Hops also add a distinctive flavor to the rural landscapes of the small areas in which they are grown. A hopyard is a thicket of poles between which wires are stretched tautly some fifteen to eighteen feet above the ground. The vines are trained to grow up strings that are attached to the wires each spring and taken down in the fall when the hops are picked (fig. 11.32).

After they have been picked the green hops must be dried in kilns in picturesque oast-houses, squat cylindrical or square structures with steep conical or pyramidal roofs. An oast-house has no windows and only a small door for access to a drying room on the upper level above a stove room on the ground floor. The peak of the roof is capped by a revolving

Fig. 11.28. A hayrack with horizontal wooden crosspieces in the eastern Alps. Some farmers attached such *Chischnern* to the sides of their barns, and these became the ancestors of the forebays on Pennsylvania German barns.

Fig. 11.29.   Bales of alfalfa hay in the Imperial Valley of southern California.

cowl, or ventilator, which has a weather vane to keep it pointed into the wind at all times, thus creating a draft for the kiln (fig. 11.33).

Ginseng

Ginseng is another small-acreage crop that has an extraordinary impact on the landscape. Ninety-five percent of the ginseng cultivated in the United States is grown under special shades in a twenty-mile radius west of Wausau, Wisconsin (fig. 11.34). Ginseng properly qualifies as a mystery crop because fewer than fifteen hundred farmers grow it, and they guard their secrets jealously. It is said that the only way to learn how to grow ginseng is to inherit the knowledge or to marry it, and there is no way an outsider can get into the business. Growers refuse to reveal their secrets even to plant scientists; in a sense they have been shooting themselves in the foot, because scientists might be able to help them cope with some of the problems that bedevil them.

The growers might retort that they do not want or need any help because they are already netting twenty thousand dollars or more a year from a single acre of ginseng, and they are properly concerned that the revelation of their secrets might encourage overproduction, which could severely depress the price they receive for the crop. A thousand acres of harvested ginseng probably can produce all of the cultivated root they can sell.

Ginseng is a bushy plant twelve to twenty-four inches tall that grows wild on the shady floors of hardwood forests in eastern North America. The fleshy root is highly prized by the Chinese, who consider it a remedy for most human ills, and they are willing to pay extravagant prices for it. Americans are a bit more skeptical of its powers. Jake Plott, who has been hunting and dealing in wild ginseng near Blairsville, Georgia, for more than half a century, said "I'll tell you the truth, there ain't a thing in the world it's good for except to make money."

Jesuit priests near Montreal began hunting wild ginseng for export to China as early as 1700, and ever since then, collectors have been scouring the woods in search of the valuable root. It has become increasingly scarce as a result of overharvest and the clearance of the sheltering forest, and the wild plant has been classified as a threatened species under the Endangered Species Act. Commercial growers must have a license to sell their root.

Ginseng is a cantankerous plant to cultivate, but around 1900 a family near Wausau learned to grow it by transplanting wild plants onto the farm where they raised mink and silver foxes. Only a few hundred acres were

Fig. 11.30. Orange groves in central Florida. Sloping land provides good air drainage, and lakes moderate the air temperature to reduce the danger of frost damage.

Fig. 11.31. Vineyards in Alsace.

cultivated until after World War II, but production increased rapidly during the farm crisis of the 1980s, when farmers were thrashing about for alternative sources of income, and in the early 1990s some three thousand acres were devoted to the crop.

The distinctive feature of the ginseng landscape is the canopies that shade the "gardens" to simulate a forest-floor environment (fig. 11.35). These canopies, of closely spaced wooden laths or of black plastic with alternating light and dark strips, are supported seven feet above the ground on wooden posts set in twelve-foot grids for laths or twenty-four-foot grids for plastic. They must be removed in winter to reduce snow damage, and erected again in the second week of May after the threat of a late snowfall has passed.

Beneath the canopies, farmers form raised beds a foot high and four feet wide, with walkways (that double in brass as drainage channels after heavy rains) a foot or so wide between them. In the fall the seeds are planted eight inches apart and mulched with three or four inches of oat straw or sawdust. Commercial fertilizers or manure would reduce the resemblance of cultivated ginseng to the wild root, and thus reduce its value, so the farmer must rely on tree leaves, old hardwood sawdust, or ground-up rotted hardwood for fertilizer.

Fig. 11.32. An irrigated hopyard in the Yakima Valley of Washington.

The roots do not mature for four long years, during which the farmer must weed the garden regularly and protect it against an awesome variety of diseases and insects. In late June, blossoms form on three- and four-year-old plants, and workers pick the seed-bearing fruit by hand when it turns bright crimson in late September. They rot the pulp to separate the seeds, which remain dormant for eighteen months, so the farmer must store them in damp sand until they are ready to be planted.

The roots finally become ripe in September or October of the fourth year, which is no time to be near a ginseng garden; the roots fetch forty to sixty dollars a pound, and many farmers protect their ripe crops with fierce guard dogs and even automatic weapons. They dig the roots with machines, but pick them up by hand. A good root is four inches long and up to an inch thick, and the usual yield is two to three thousand pounds per acre. The roots are washed and then dried for two to six weeks, and then they are ready for sale to Chinese buyers who visit the farms to haggle for them and export them to Hong Kong.

Few growers dig more than an acre a year because each acre requires six hundred hours of work each year for shade maintenance, weeding, seed picking, digging, washing, and drying. Growers have made a major capital investment before they harvest their first crop, and it takes them six to ten

Fig. 11.33.   Oast-houses for drying hops.

years to break even, but then they really start to rake in the money. There is one final quirk. It is difficult, perhaps even impossible, to grow more than a single crop of ginseng on a given piece of ground, either because of disease organisms in the soil or because the plant is autotoxic, and eventually growers have to lease or buy new land for their gardens.

Cranberries

One-fifth of the nation's cranberries are produced in three counties southwest of Wisconsin Rapids, only fifty miles from the ginseng patch west of Wausau (see fig. 11.34). Compared to ginseng cultivation, cranberry production looks simple—but only when compared to ginseng cultivation. Cranberries grow on low, vinelike, woody perennial bushes with narrow green leaves that turn a dull reddish brown during the dormant season. The long-lived plants, which continue to produce a crop of berries each year for decades, grew wild on the acid soils of peat bogs in the glaciated areas of New England and the Midwest. The Indians collected the berries for medicinal purposes because they are an excellent source of Vitamin C. They keep well, and sailing ships used to carry barrels of cranberries to help the sailors ward off scurvy on long ocean voyages. The British navy used lime juice, but American sailors munched on cranberries.

Farmers on Cape Cod started growing cranberries on artificial bogs around 1820, and production had spread to New Jersey by 1835 and to central Wisconsin by 1853. Blueberries thrive on the same kind of acid soil that cranberries like, and they have largely replaced cranberries in New Jersey, but Massachusetts and Wisconsin still produce more than three-quarters of the nation's cranberries. The growers on Cape Cod noticed that the best berries grew at the edges of the bogs, where wind-driven sand from the neighboring dunes covered the stemmy old plant growth, and they learned to spread an inch of sand on their bogs every three or four years to rejuvenate the bogs and stimulate new growth.

Cranberry growers create their bogs by clearing off the vegetation and leveling poorly drained areas that are underlain by impervious subsoil

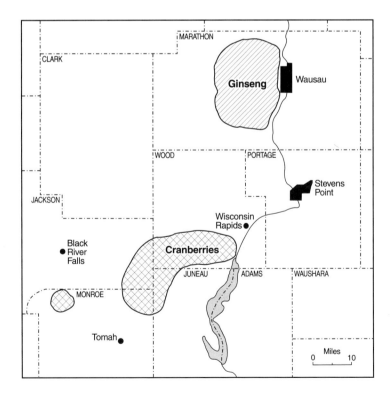

Fig. 11.34. The ginseng and cranberry areas of central Wisconsin.

Fig. 11.35. Ginseng gardens are shaded by canopies of wooden laths to simulate a forest-floor environment.

hardpans. The minimum size for a commercial bog is about thirty acres, which is divided into management units of two to four acres (fig. 11.36). Each unit is enclosed by a grass-covered earthen dam twelve feet wide with an access road on top. The top of the dam must be two feet higher than the maximum water level in the unit. The dams have gates that the farmer can open or close to control the water level in the unit. Each unit has drainage ditches two feet deep on all four sides just inside the dam, and shallower ditches across the center, because the farmer must be able to flood and drain the unit quickly.

Growers spread a two- to six-inch layer of clean sharp sand on top of the peat in a new bog, and sand it again as appropriate. They propagate cranberries vegetatively by cutting prunings from older bogs into six- to eight-inch lengths and planting them in April or May. They can expect their first commercial crop in the fourth autumn after they plant, and thereafter the bog is essentially permanent.

The ripe berries are harvested in early October before the first frost. In the early days they were picked by hand; then they were combed from the vines with hand scoops, and the growers brought in gangs of migrant workers for the harvest. After World War I some growers got the idea of water

Fig. 11.36. Earthen dikes divide cranberry bogs into management units of two to four acres.

harvesting. They flooded a unit to lift the vines, and they could easily rake off the berries as they floated near the surface. Then they began building their own water-harvesting equipment. On a small tractor they mounted five reels, like large rotary lawn mowers but with rods rather than blades. The revolving rods beat the berries loose from the vines, and they float to the surface where they can be skimmed off. A flooded cranberry bog whose surface is crowded with floating red cranberries is a spectacular sight.

Cranberry bogs are in low areas subject to air drainage, and the growers flood them from November to April to reduce the risk of frost damage to the plants. They must have an abundant supply of pure water from streams or wells for irrigation, for harvesting, and for the winter flood, and each acre of bog needs two acres of storage reservoir. Legal restrictions on the use of water and wetlands are a major problem for the growers, and they will make the development of new bogs increasingly difficult.

Rice

The rice plant is a native of marshy areas in the tropics, and for most of the growing season it must be in at least an inch and a half, and no more

Fig. 11.37. Control structures on a lateral irrigation ditch.

than four inches, of standing but not stagnant water. Rice farmers need land that is flat enough to flood, but with enough slope that it can be drained fairly quickly. They divide their fields into "cuts" by making levees fifty to a hundred feet apart that snake along the contours. The contour interval between levees is two tenths of a foot on level land and three tenths of a foot on "steep-fall" land. The levees are eighteen inches high to hold water in the cut, and nine to ten feet wide so machinery can be driven across them easily. Each levee has a gate that can be opened to fill the next-lower cut when its own cut has been flooded.

Every acre of rice grown in the United States is irrigated. Rice farmers pump their irrigation water from streams or wells. Water straight from a well is so cold that it harms the plants, and it must be held in an open reservoir until it has warmed up. Some rice farmers stock their reservoirs with catfish and bass. Open ditches or buried pipes carry irrigation water to the fields, and relifts at the lower ends of the fields pump it back to the reservoir so it can be used again. Every rice farm has a dirt airstrip, because the farmers need crop-dusting airplanes to spray the agrichemicals that will kill weeds and insects and control diseases when their fields are flooded and they cannot get into them with conventional machinery.

RESHAPING THE LAND FOR WATER

Rice and cranberry growers are not the only farmers who have re-shaped the land to meet the needs of their crops and to manage the movement of water to, across, through, and from it. In much of the western United States, crop production would not be possible without water for irrigation. Most major irrigation systems have massive dams and large storage reservoirs, or deep wells and powerful pumps, with a network of canals, laterals, and ditches to get water to individual fields (fig. 11.37). Traditional gravity-flow systems required the fields to be as flat as a tabletop, with just enough slope so water would flow, but not enough to let it flow so rapidly that it would erode the soil. Farmers leveled their fields with large scrapers called land planes, which shaved off the bumps and filled in the depressions when tractors pulled them across a field.

The development of sprinkler irrigation systems has enabled farmers to irrigate land without having to level it. Center-pivot sprinkler irrigation systems, which came into general use in the 1960s, irrigate circular "fields" that stand out quite strikingly in the rectilinear American landscape (fig. 11.38).

Fig. 11.38. Circular fields produced by center-pivot irrigation systems.

Fig. 11.39. Alternating strips of wheat and fallow land are the hallmark of dry-farming areas on the Great Plains.

In the drier western parts of the Great Plains the circles produced by center-pivot systems may be close neighbors to the striped fields associated with dry farming. Dry farmers alternate strips of wheat and fallow land, and use the rain that falls on this year's fallow strip to grow next year's wheat crop (fig. 11.39).

Paradoxically, areas that are irrigated must also be drained to carry away the alkaline groundwater that would otherwise poison crops (fig. 11.40). Drainage on a large scale has been necessary too in the humid East, where extensive areas, including some of the most productive parts of the Corn Belt, could not have been cultivated without the development of drainage systems to remove excess water. It is all too easy to overlook the importance of drainage; it makes adjacent areas look more alike rather than more distinctive, and drainage works, apart from open ditches, are not readily visible, although the herringbone pattern of buried tile drains may be strikingly apparent from the air during a wet spring.

Finally, in areas of steep slopes farmers must reshape their land to prevent rainwater from leaving it too rapidly and taking some of the topsoil with it (fig. 11.41). The erosive ability of running water is a product of its velocity, which in turn is a product of the slope, so farmers with steep land must do what they can to slow down the speed of rainfall runoff. Some drape their fields along the contours; others heap up low earthen terraces along the contours to reduce runoff. The low earthen terraces that snake through fields, forests, and pastures alike in the modern South are useful indicators of the former extent of cultivated land, because they were originally constructed to reduce the danger of erosion in cotton fields (see fig. 1.7).

In some densely populated parts of the world, where cultivable land is scarce and precious, entire mountainsides have been transformed into

Fig. 11.40. Drainage outfall in an irrigated area. The saturated soil is dark. The light patches on the dark soil are alkaline salts deposited when groundwater evaporates.

Fig. 11.41. A steep hillside terraced for vineyards in the Moselle Valley of Germany.

Fig. 11.42. A cotton gin.

spectacular steplike terraces on which crops can be grown. A prodigious expenditure of cheap human labor is necessary to build the stone retaining walls and to carry baskets of earth up the steep slopes to fill in the areas behind them, and the returns for this effort may be pitifully small, but they might be just enough to stave off starvation.

OFF-FARM STRUCTURES

Most farm products must be processed in some way before they are ready for sale or shipment to consumers. Some can be processed on the farm, but most agricultural areas need structures where the products of many farms can be collected and processed, and from which they can be shipped. Often these processing plants also perform a marketing function because they buy the products they process. Cotton farmers sell their crop to the cotton gin (fig. 11.42), dairy farmers sell their milk to the creamery, peanut farmers sell their nuts to the drying warehouse, sugar producers sell their cane or beets to the sugar mill, vegetable farmers sell their crops to the cannery, and vineyard owners sell their grapes to the winery. Each of these processing structures is distinctive, and each is characteristic of the small central places that serve the agricultural area where the particular commodity is produced.

Distinctive structures that buy and ship but do not process agricultural commodities are characteristic features of the central places of other agricultural regions. Market towns in tobacco-producing areas have cavernous sales warehouses, their roofs studded with rows of skylights to illuminate the sales floors where the leaf is auctioned (fig. 11.43). Many small towns in livestock- and truck-farming areas have local auction barns, where sales are held in season. Central stockyards, where farmers sell their fat animals to meat-packing companies, generally have been in larger places than these other sales structures, but all are as much a part of the distinctive landscape of an agricultural region as are its barns and fences.

Grain elevators are a fine example of the kinds of off-farm structures that are essential for agricultural regions. Nearly every whistle stop on every railroad line running through every grain-farming region (corn and soybeans from Iowa to Ohio, wheat on the Great Plains, rice in Arkansas and southeastern Texas) seems to boast its own trackside battery of elevators, which are minor cogs in a highly organized network for collecting grain from farmers and getting it to markets at home and abroad.

Fig. 11.43. A tobacco sales barn with rows of skylights on its roof.

Fig. 11.44. Grain elevators.

Farmers dump their grain into a pit at the base of the elevator. A vertical conveyor lifts the grain to an overhead system of belts and chutes that direct it into vertical storage bins, from which it can be loaded into railroad cars, trucks, or barges for shipment.

In the late nineteenth century, subsidiaries of railroad companies built three- or four-story wood-frame elevators at intervals of about five miles along the railroad lines that were pushing into grain-producing areas. Small "country" elevators bought grain from local farmers and shipped it to much larger terminal elevators (fig. 11.44). After 1900 most new elevators were cylindrical concrete silos, and many were built or bought by co-operatives of farmers, who coveted the profits the elevator companies had been making on their grain.

Increased grain yields since 1950 have forced many small elevators to add new metal grain bins to store the increased amounts of grain that local farmers are producing. Technological changes in grain storage and transportation have made many country elevators obsolete, and they are no longer used, but still they dominate the skylines of small towns in grain-producing regions, silent reminders of days gone by.

12 *Farm Size and Farm Tenure*

Any area is a mosaic of many individual properties. In rural areas most of these properties are farms or ranches, and the way their land is used will determine the appearance of the landscape. In order to read the landscape aright we must understand the multitudes of independent decisions made by the hosts of individuals who decide how the land is used. We need to identify the individuals who have made these decisions, and we must try to understand why they made them as they did. We need to know who owns the land, who controls it if it has been leased, and what factors influence their decisions.

THE OPERATING UNIT

The student of the rural landscape must be sensitive to the fundamental distinction between ownership units and operating units, because the person who owns the land is not necessarily the person who farms it. This distinction hardly mattered back in the days of small farms, when most farmers owned all the land they farmed and farmed all the land they owned, but it has become increasingly important as farms have gotten larger and the functions of ownership and operation have become more separate. The owner holds legal title to a property, but the operator does all the work on it or directly supervises the work.

On any given farm the functions of owner and operator may be combined in the same person, although they need not be. The person who owns a farm may operate it in person, as owner-operator; the owner may entrust its operation to a manager, overseer, or bailiff; or the owner may lease some or all of it to another person as tenant. The person who operates a farm may be the full-owner of all the land in the operation, a part-owner who owns part of the land and rents the rest of it from someone else, or a tenant who rents all the land. In 1992, full-owners operated 31 percent of all farmland in the United States, part-owners operated 56 percent, and tenants operated only 13 percent.

As a general rule, owners who lease their farms to others are responsible for the provision and maintenance of their permanent structures, such as barns and fences, and thus the owners have a significant impact on the look of the land even though they are not actually farming it. Owners may also stipulate the kinds of livestock the tenants may keep, the kinds of crops they may grow, and perhaps even the rotation in which they must grow them. Tenants do not have to accept leases they consider too restrictive, and they are perfectly free to shop around for leases they like better. They may also take the initiative in encouraging the owners to invest in new permanent structures, so they too can play an important role in influencing the look of the land.

The boundaries of ownership units are more permanent than the boundaries of operating units because the former are defined by legal title in the public records, and they can be changed only at the expense of the services of lawyers, title abstractors, registrars of deeds, and other functionaries. The boundaries of individual properties are easy to identify and map, but the compilation of ownership maps from legal records for an area of any size is a truly tedious task. In many counties in the United States this task is made less onerous by the availability of plat books that show the boundaries and names of the owners of all properties. Plat books are a useful source of information about ownership units.

A map of ownership units, unfortunately, is not a map of operating units, because owners may have rented part or all of their land to others. The boundaries of operating units are even more difficult to identify and map than the boundaries of properties; tenancy arrangements may be defined by a formal contract or lease, or by an informal oral agreement and a handshake. They expire automatically after a stipulated period, sometimes only a year, and then they are subject to renegotiation and change. Commonly the boundaries of operating units can be identified only by interviewing each and every person who owns or leases land in an area. In the United States the county offices of the Agricultural Stabilization and Conservation Service have information about operating units, but ASCS personnel are understandably reluctant to release this privileged information without explicit permission from the individual operators, and it is not easy to use.

Despite the difficulty of identifying and mapping them, operating units are critically important to anyone who wishes to understand the rural land-scape, because the operator makes many, if not most, of the decisions that determine how the land will be used and how it will look. The operator decides what crops will be grown (and in what rotation), what livestock will be kept, and what type of farming will be practiced; these facts are rec-ognized by most censuses of agriculture, which collect and publish data on operating units rather than ownership units. The United States Census of Agriculture, for example, is a census of operating units rather than owner-ship units, and all data are reported for the county that contains the head-quarters of the operation, even though some of its fields might be in another county or even in another state. This practice explains why the total acreage of farmland reported for some Iowa counties is greater than the area of the counties.

FIELD PATTERNS

One of the most striking manifestations of farm operating units on the landscape, whether as seen from the air or viewed in aerial photographs, is found in their subdivision into fields. Although field patterns in most areas are remarkably complex, each and every field has its present size and shape because at one time that particular size and shape made sense to somebody. As a general rule, livestock farmers prefer small, squarish fields because they permit better control of grazing and breeding, whereas crop farmers are more inclined toward long, narrow fields, which are better suited to the use of machinery (fig. 12.1). As an even more general rule, large farms are divided into large fields and small farms into small ones, because there is an upper limit (often related to crop rotation) to the number of fields one operator can manage efficiently. Few farms, no matter how large they are, have more than eight or ten fields, and not many have fewer than three.

In most areas the field pattern is closely related to the system under which the land was surveyed. Within a given system of land survey, varia-tions in the form of the land surface are probably the primary determinant of variations in field patterns (see fig. 8.11). Striking examples of the influ-ence of landforms are fields in which the crops are planted in strips along the contour at right angles to the natural slope of the land (fig. 12.2).

Field patterns may also be influenced by farming systems. In dry-farm-ing areas in the West, for example, alternating strips of land are cultivated and left fallow, which gives the landscape a distinctive striped appearance.

In Indiana I found small, square fields in an area of mixed corn-hog farming and long, narrow fields in an area of highly mechanized cash-grain farming (see fig. 9.20). Farmers in the cash-grain area were removing fences and enlarging their fields so that the rows in each field could be as long as possible to minimize the amount of time wasted in turning their machinery around at the ends of rows. Fifteen years later the corn-hog area had also shifted to cash-grain farming, and the farmers were modifying their fields in precisely the same way.

The layout of fields often is even more conservative, and changes even more slowly, than farming practices in general. In Europe, where geographers have studied them much more intensively, field patterns seem to be more persistent than they are in the United States, where fields are merely convenient subdivisions of farm operating units. In Europe many individual fields are stabilized by such difficult-to-remove boundaries as hedgerows, earthen banks, and stone walls.

THE SIZE OF FARMS

The size of a farm is one of the most important considerations that influences the way its land will be used, because the acreage of land with which farmers have to work may either constrain or enhance their options when they make decisions about their operations. A small farm, for example, must produce a greater return per acre to generate an acceptable level of living, and small farms generally are cultivated more intensively and require more labor per acre than larger farms do. Farmers with small acreages must try to produce more-valuable products, or greater quantities per acre, or both, than farmers with larger acreages.

The efficient operation of modern farm machinery requires a minimum acreage of land, and farmers who have less than this minimum cannot afford the machinery they need because its fixed cost per acre is too high. Large farms, on the other hand, require greater management skill because the effects of mistakes are magnified and a large farm can lose money faster than a small farm.

Acreage

Acreage is the customary measure of farm size, but in the United States as a whole the average size of farms in acres is a misleading statistic. It is so

Fig. 12.1. Long, narrow fields reduce the amount of time wasted in turning machinery.

heavily weighted by huge landholdings in the drier parts of the West, where carrying capacities are low and ranchers talk in sections (square miles) rather than in acres. In 1987, for example, Esmeralda County, Nevada, led the nation with an average farm size of 64,244 acres, or 100.38 square miles. Six of these farms put together would have been larger than an average Iowa county, and ten of them would have blanketed the entire state of Rhode Island.

Even in the humid East the average size of farms can be misleading because some crops, such as nursery products, tobacco, fruits and berries, vegetables, Irish potatoes, and cotton, generate high gross sales per acre and are conducive to small but intensively cultivated acreages, whereas field crops that generate low gross sales per acre need relatively large acreages. For instance, one acre of tobacco generates the same average return as two acres of vegetables or fifteen acres of corn.

Confined-feeding operations for poultry, hogs, and cattle also can generate large gross sales per acre, although their net returns are reduced by heavy expenses for feed and animals. For example, a single thirty-two-by-four-hundred-foot poultry house, which occupies less than a third of an acre, can put through six flocks of sixteen thousand birds a year, or a total

Fig. 12.2. Fields are laid out in strips along the contours to reduce the risk of soil erosion.

of ninety-six thousand birds, which would have had a gross sales value of approximately fifty thousand dollars in 1992. Little wonder, then, that tobacco- and poultry-producing areas have large numbers of small-acreage farms. Small-acreage farms are also numerous in perimetropolitan areas that have intensive nurseries, greenhouses, and truck farms, as well as many hobby farms whose owners are pouring money into them instead of making money from them.

Four states in the heart of the Corn Belt, which have a large share of the nation's family-sized farms, show how dramatically the average acreage of farms has increased in the United States since World War II (fig. 12.3). Between 1949 and 1992 the average acreage of farms in Iowa, Illinois, Indiana, and Ohio more than doubled, after half a century during which it had hardly changed. The largest farms have consistently been in the western part of the area, which was settled later when improved technology enabled farmers to handle greater acreages.

The number of farms must decrease when their average size increases, and the number of farms in the four Corn Belt states shrank from three-quarters of a million in 1949 to only one-third of a million in 1992.

Number

Official census statistics on the acreage, number, and other attributes of farms in the United States are invaluable, but they can also be extremely misleading for anyone who has not learned how to use them correctly. For example, the official census definition of a farm is so permissive that it includes large numbers of operations that can best be described as nonfarm farms—a variety of undersized, part-time, low-income, weekend, retirement, and hobby operations that could not be considered genuine farms by the wildest stretch of anyone's imagination.

The official definition of a farm has been changed nine times since 1850, but it has always remained permissive. Since 1974 the agricultural census has defined a farm as "any place from which $1,000 or more of agri-

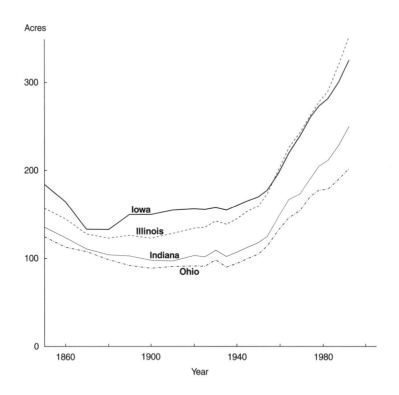

Fig. 12.3. The average size of farm in four Corn Belt states remained fairly stable between 1860 and 1940, but it has increased rapidly since 1950. The trends in all four states have been remarkably similar.

cultural products were produced and sold or normally would have been sold during the census year." Few people could manage to live on a gross income of only a thousand dollars a year, but more than one-tenth of the so-called farms counted in the 1987 agricultural census actually sold less than that amount, and more than one-third sold less than five thousand dollars' worth of farm products.

This extremely permissive census definition of a farm should not be changed, however, because it allows the census to provide complete and detailed information about farming in the United States. It is better to collect too much information than too little; we can always ignore data that we do not need, but we cannot hope to use data that have not been collected. Furthermore, human beings live on each of these "farms," no matter how uneconomic they may seem. Some of these people derive part of their livelihood from their farms, and they have an impact on the way the land is used and the way it looks.

Nevertheless, users of census data should not accept them at face value without understanding that nonfarm farms account for a significant portion of all operations. For example, census reports show that the total number of farms in the United States dropped from almost seven million in 1934, the peak year, to fewer than two million in 1992. The loss of almost five million farms in half a century has unduly alarmed those who have failed to realize that most of the "lost" farms were not viable agricultural enterprises.

Even in 1992 at least half of the farms included in the census were nonfarm farms. Probably the best single indicator of the number of nonfarm farms is provided by the responses to the census question, "At which occupation did the operator spend the majority (50 percent or more) of his/her worktime in 1992, Farming/ranching or Other?" The proportion of farm operators who said "Other" has remained surprisingly constant near 45 percent since this question was first asked in the 1974 census, even though the nation has lost some two hundred thousand farms at each subsequent census. In county after county in 1992 the number of farm operators with "Other" listed as the principal occupation, the number who worked "off farm" one hundred days or more, and the number of farms with sales of less than five thousand dollars were so nearly similar that these measures are interchangeable indexes of nonfarm farms.

It is tough to tell whether the distribution of nonfarm farms in the United States is more closely related to the pull of off-farm employment opportunities or to the push of farming systems that do not employ their operators for the full day, such as poultry farming, or for the full year, such as tobacco and cash-grain farming. For example, the fewest nonfarm farms are in dairy-farming areas, which demand a full-time effort all year round, but appreciable numbers of nonfarm farmers are employed in forestry in the Pacific Northwest, in mining in Appalachia and in east Texas, and in nearby towns and cities throughout the nation.

Sales

An alternative measure of farm size is the gross annual value of sales of farm products. The gross value of sales, like any other indicator, can be misleading; some crops and animals are more valuable than others, and some types of farm produce lower net returns than others that have the same gross value of sales because they have to buy more of their inputs. Most of the income of tobacco farmers, for example, goes into their pockets, but much of the gross income of poultry farmers goes to pay for the feed they have to buy for their birds.

The list of possible farm expenses is awesome. On most farms it includes the cost of seed, fertilizer, pesticides, and other necessary agrichemicals; machinery, gas, and oil; livestock and feed; taxes, insurance, interest, and bank charges; telephone and electricity; and the maintenance and repair of buildings, fences, and the other fixed equipment of the farm. Most of the gross income of the farm has to go to pay its expenses, and farmers rarely can use more than 5 to 10 percent of the money that passes through their hands to pay for family living expenses.

The minimal gross annual value of sales of farm products needed to provide an adequate level of living for the farm family, like just about everything else, has kept going up since World War II. The rule of thumb says that it was probably around five thousand dollars before 1960, ten thousand dollars in the 1960s, forty thousand dollars in the 1970s, and close to a hundred thousand dollars in the early 1990s, but such figures are no more than rough guesses. We do know, however, that the farms that exceed these minimal values have produced 85 to 90 percent of the nation's farm products, and smaller farms have produced a mere pittance. On September 21, 1994, the *New York Times* stated that farms with sales of less than a hundred thousand dollars a year are part-time or hobby farms, and in 1993 the Federal Reserve Bank of Kansas City said that many lenders believed that a farm had to have sales of five hundred thousand dollars or more to be financially viable.

In 1992, for example, 1.4 million farms in the United States sold less than fifty thousand dollars' worth of farm products, and only 47,000 farms sold more than five hundred thousand dollars' worth. The smallest 73 percent of all farms produced only nine percent of all farm products, while the largest 2.4 percent of all farms produced 46 percent of all products (fig. 12.4). These numbers reinforce the idea that many of the farms reported in the census could not possibly support a family at any acceptable level of living, and most of such farms will eventually be abandoned or incorporated into other units. The United States could easily "lose" another mil-

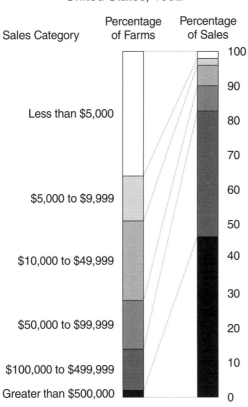

Fig. 12.4.   In 1992 three-quarters of all farms in the United States produced only one-tenth of all farm products, and one-tenth of all farms produced three-quarters of all farm products.

lion farms without any reduction in its ability to feed and clothe itself, and probably half of the million that remained would still be nonfarm farms.

THE FAMILY FARM

"Family farm" is a shibboleth. The ideal of the family farm has taken its place close to God, motherhood, and *The Star-Spangled Banner* in the national pantheon. The ideology of the family farm is an amalgam of many of the traditional frontier values: the self-made man is better than the wellborn one, so every man should have enough land to enable him to prove himself; he should own his land in fee simple, unencumbered by debts; his farm should be large enough to provide him with a decent livelihood, so that he can accept responsibility for his economic security and that of his family; he and his family should provide a substantial share of the labor on it, but they should receive a fair return for their efforts (the prices farmers get should bear a reasonable relationship to the prices farmers have to pay); their income should be related to their diligence, because hard work is a virtue; and the worth of a man may be measured by the amount of money he makes.

Generations of politicians have bloviated on and on about the virtues of individual farm ownership as one of the cornerstones of a free and democratic society, and an enormous amount of drivel has been perpetrated about family farms, much of it obviously by people who have not the foggiest idea clue as to what a modern family farm really is, or how rapidly family farms have been changing. Many of these sentimental ideologues would flatly refuse to live and labor on the kinds of hardscrabble operations they have so fondly romanticized.

Some people still cherish the notion that a family farm is a self-sufficient operation with a few acres and a milk cow and some pigs and a few chickens and a garden patch. The only people who can actually afford to live on such places today are those who have a guaranteed source of outside income, and the most important structures on their farms are the mailboxes in which they receive their checks.

A typical modern family farm is an agricultural operating unit that provides an acceptable level of living in return for the full-time labor of a parent and child, most often a father and son, perhaps with a hired hand at those stages of the demographic cycle when the son is too young to be of much help or when the father is too old. On many farms the farm wife has been called on to fill the role of the hired hand because good workers are hard to find, and brains have increasingly replaced brawn as the requisite for successful modern farming. Often when she has not been needed on the farm the farm wife has taken a job in town to help maintain a steady flow of cash for the family farm business.

Neither size nor ownership is stipulated in this definition of a family farm, nor should either be. The amount of land that will provide a reasonable farm income under intensive cultivation in an irrigated area in the West would hardly be enough to keep a single cow alive on adjacent sagebrush-covered rangeland. Livestock ranches and highly mechanized grain farms both require and can use more land than farms that specialize in labor-intensive vegetables, fruit, tobacco, poultry, and dairy products, which demand individual attention and considerable hand labor. Recent changes in agricultural technology not only enable individual farmers to care for a greater amount of land, but also require them to do so if they are to receive an adequate return for their efforts. No matter what the farming system, the amount of land that was large enough for a family farm a generation ago has become, or is becoming, too small, and the size of the operating unit must be enlarged.

Some people have equated family farms with small farms, and they have jumped to a whole suite of erroneous conclusions: if family farms are small, then small farms must be good, so large farms must therefore be bad, and large farms cannot be family farms. They snarl the term *agribusiness* as though it were a dirty word. Some of them even assume that large farms must be owned by corporations, so corporate farms must be exceptionally bad.

Highly specialized megafarms in Florida and in the irrigated oases of the West have indeed been incorporated because of their sheer size, and they are run by professional managers, but many family farms in the East have been incorporated in order to keep them in the family by ensuring the successful intergenerational transfer of assets. A family farm must be large enough to support the families of two workers, a parent and a child. The family-held corporation or partnership does not have to sell off part of the farm to pay inheritance taxes when one member of the family dies.

A successful modern family farm is a business, so I defined the family farm in business terms in order to use census data to map them. I used the gross annual value of sales of farm products as a measure of farm business size. In 1982 635,000 farms in the United States sold more than forty thousand dollars' worth of farm products, which is a conservative cutoff value.

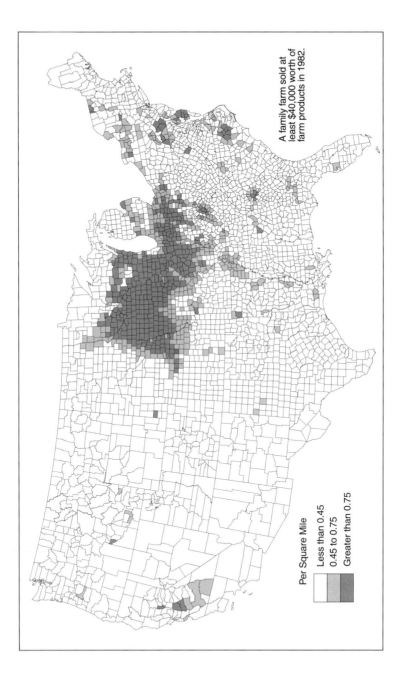

Fig. 12.5. Family farms that sold at least forty thousand dollars' worth of farm products in 1982 were heavily concentrated in the agricultural heartland in the Midwest.

These family farms were heavily concentrated in the nation's agricultural heartland in the Middle West (fig. 12.5).

Farming in the Midwest has undergone a revolution in recent years, and few Americans know much about it or how it has changed. Even those who grew up on family farms before 1960 will have a tough time understanding them today if they have not been going back for regular visits. The number of people who live on farms has been dropping at an astonishing rate. Each of the four Corn Belt states has lost around one hundred thousand farm people at each census since 1940, and except for Iowa, with a quarter of a million, none of them had more than about two hundred thousand farm people left in 1990 (fig. 12.6).

FARM ENLARGEMENT

Successful modern family farmers have had to keep enlarging their farm businesses in order to stay in business. The cost of everything they have needed has been rising steadily, but the prices they have received for their crops and animals have not kept pace with their costs. Their profit margins per unit have been shaved razor-thin, and the only way the farmers have been able to keep their heads above water has been to produce more units. The farm business, like the grocery business, has changed from low volumes with good profits per unit to huge volumes with minuscule per-unit profits. The high-volume family farm business has replaced the traditional small family farm, just as the supermarket has replaced the traditional corner grocery store.

Farmers have increased their volume of production by using new technology that has enabled them to produce more from each acre. Plant breeders have given them improved varieties of crops that are tailored to each and every ecological niche, and their yields have doubled or even tripled. Chemists have given them better fertilizers to help them take advantage of the full potential of the new varieties, and a whole arsenal of pesticides to protect their crops against weeds, insects, and diseases.

Engineers have given them bigger, more-sophisticated, and more-expensive machinery to enable them to cope with their increased yields and larger acreages. Most modern farms have at least four tractors of various sizes to handle the myriad jobs that have to be done on the farm. Self-propelled combines, which harvest and shell their crops in one single, swift operation, have forced farmers to invest in new cylindrical storage bins of

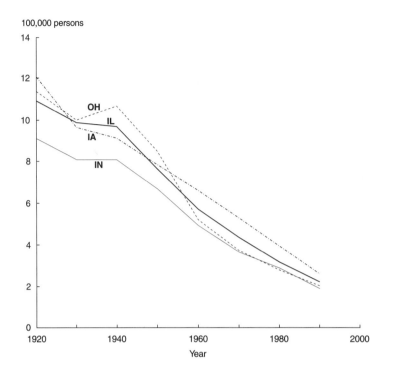

Fig. 12.6. The farm population has been declining steadily and consistently in each of four Corn Belt states.

gleaming corrugated metal, and in facilities for drying the crops when they come in from the field.

Many modern farm machines are equipped with global positioning systems (GPS) that tell their operators precisely where they are as they lumber across the land, and the operators use them to adjust their rates of seeding and applications of fertilizer and pesticides to subtle variations in the physical character of the soil.

A modern family farm business is run just like any other business, and it is so complex that the farm office must be as well equipped with computers, fax machines, and the like as any business office in town. The new technology costs money, lots of it, so modern family farmers have had to become skillful money managers who spend hours staring at their computer screens. They have had to learn to specialize in doing what they do best, and they have had to eliminate those activities that their computers tell them are less profitable. They have had to learn to manage money just as adroitly as they manage crops and livestock and machines. They used to think in terms of hundreds of acres and dollars, but now they must think in thousands, and ulcers have replaced blisters as one of their principal occupational hazards. A successful modern family farm is a business with a larger capital investment and a greater gross volume of sales than most of the businesses in the small towns that serve them.

PART-OWNERSHIP

Modern family farmers have also had to farm more acres in order to increase their volume of production. The ideal way to enlarge a farm business would be to buy more land, but many farmers are in no position to do so. Like most of the rest of us, few farmers have as much money as they would like to have, or even as much as they need, and the only way they can get more land is to rent it rather than to buy it. Part-ownership has become the customary strategy for farm enlargement.

Moreover, farmers may have to rent land even if they can afford to buy it, because their neighbors are unwilling to sell it. Some of these neighbors may think of their land as an investment that is too valuable to part with, or they may simply enjoy the prestige of owning it. Others may have a sentimental attachment to the old home place, which has been in the family such a long time. Farmers can understand these attitudes, but cold-blooded city economists often are baffled, irritated, and impatient when they discover that many rural landowners have a strong, deep-down-in-the-bones feeling that it is somehow wrong, perhaps even sinful, to part with a piece of land for mere money. Perhaps the economists are the ones who are irrational.

The farm crisis of the early 1980s illustrated the folly of paying too much for farmland. The 1970s were good years for farmers, and the price of farmland kept rising so rapidly and so steadily that nonfarm speculators such as investment funds and insurance companies decided that it was a good investment, and they drove the price to unprecedented heights. They got burned when the price of farmland cascaded in the early 1980s, and their screams of anguish garnered considerable publicity for a farm crisis that was largely of their own making. Those farmers who had expanded unwisely and bought overpriced land at exorbitant interest rates were also hurt, but the good farm managers, who had rented land instead of buying

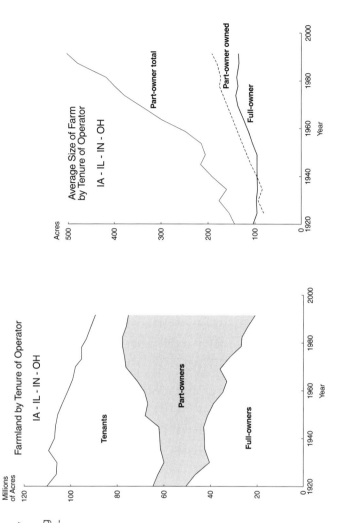

Fig. 12.7. The share of Corn Belt farmland operated by part-owners has increased steadily since 1950. Part-owners apparently have increased the size of their operations by renting land rather than by buying it, because the average acreage owned by part-owners is close to the average acreage owned by full-owners.

it, simply tightened their belts, cut their expenses, lived on their equity, and rode out the storm.

Part-ownership has been a powerful and necessary strategy for farm enlargement. In the four Corn Belt states the proportion of farmland operated by part-owners increased from less than one-quarter in 1949 to more than half in 1992, and the average size of part-owner operations more than doubled, while the average acreage of full-owner farms hardly changed (fig. 12.7). Full-owner farmers are generally older operators who are cutting back to a smaller volume of business and a smaller income as they look forward to retirement.

Livestock and fruit farmers have bought land instead of renting it because they are not willing to make necessary and expensive improvements on land they do not own, but crop farmers have rented land and used their money to buy the machines they need to work it. The agricultural census has no separate data on the acreage owned and rented by part-owner farmers between 1949 and 1978, but since then the average acreage they own, as opposed to the average acreage they operate, has been close enough to the average acreage owned by full-owners to suggest that most part-owner farmers have not bought land, but have expanded by renting it.

Farmers who already own their land, and have surplus labor, machinery, and buildings, can afford to rent smaller parcels of land, and pay higher rents for them, than tenants who must depend on the rented land for their entire income. Owners with land to lease prefer to lease it to farmers who already own some land, on the assumption that they will be more responsible and that they will not require a house to live in or any farm buildings other than a few simple shelters. A farmstead, whether or not it is occupied, holds so much attraction for the tax assessor that its owner may prefer to raze any buildings that are not necessary.

Part-ownership is also a useful strategy for keeping the better land in production in areas of mixed soils and surface features. A rectangular property in such an area may contain some good land as well as some land that is hardly worth farming. Owners may use their good land as a base of operations and put together a farm of adequate size by renting land from other owners in a similar situation, or they may take off-farm jobs and rent their better land to farmers who wish to expand. This strategy has been especially appealing in parts of Appalachia, where good and poor land are closely intermixed, and where the reluctance to sell the old family place is so strong that it is difficult to buy enough land for a farm large enough for modern mechanized operations.

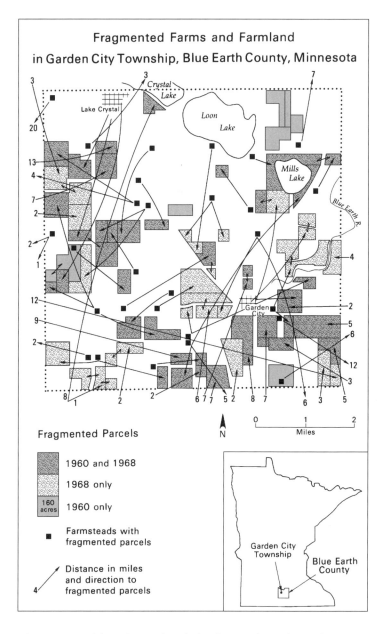

Fragmented Farms and Farmland
in Garden City Township, Blue Earth County, Minnesota

Fig. 12.8. Fragmented farms in a cash-grain farming area in southern Minnesota. Reproduced by permission from Everett G. Smith Jr., "Fragmented Farms in the United States," *Annals of the Association of American Geographers* 65 (1975): 66.

Part-ownership means that most modern family farms consist of widely scattered tracts of land; the competition for land is keen, and most farmers have to travel considerable distances to find land they can rent. In 1960 two of every five farms in nine Minnesota townships consisted of two or more separate tracts, often separated by several miles, and in 1968 more than half the farms in a Blue Earth County township consisted of scattered tracts that ranged in size from 40 to 320 acres, and were anywhere from one to twenty-two miles from the farmstead (fig. 12.8).

In 1982 the 1,200-acre Magnus family farm in southwestern Minnesota consisted of 500 acres in four tracts that Clarence and Doug owned, 540 acres in five rented tracts, and a 160-acre tract they custom-farmed, but Doug warned me that the competition for land to rent was so fierce that the map of their operation changed nearly every year (fig. 12.9).

In 1995 Wayne Schilling, whose family had farmed the land since 1858, was the last remaining dairy farmer in Woodbury, Minnesota, the most rapidly growing suburb of the Twin Cities. He had managed to put together a farm of 540 acres by renting twenty-eight separate parcels of land from as many different owners (fig. 12.10), but he realized that it was only a matter of time until all the land would be built up.

"Suitcase farmers" on the Great Plains, who live thirty miles or more from the land they farm, are an extreme example of part-owners farming widely scattered tracts of land. The Great Plains have a history of severe and prolonged droughts, and the local farmers have learned to diversify their operations as a hedge against drought and price fluctuations. Suitcase farmers from Texas, Oklahoma, and Kansas have diversified geographically by growing a single crop, wheat, in different areas stretching from Texas to Saskatchewan. They know that drought rarely hits all wheat-producing districts on the Great Plains in the same year, and by planting wheat in several different districts they have a good chance of getting a crop in at least one.

Part-owner tobacco farmers in eastern North Carolina practice geographical diversification on a smaller scale. They rent entire farms just to acquire their tobacco allotments—the right to grow a specific acreage of the crop. They continue to grow tobacco in small patches on the rented land instead of consolidating the crop in a single large field, because they know that summer precipitation in the area comes mainly in the form of thunderstorms, which are notoriously spotty. Scattering tobacco fields around the county reduces the danger that the entire crop will be wiped out by hail, parched by the failure of thundershowers to hit a particular field, or drowned by torrential showers that overload field drains.

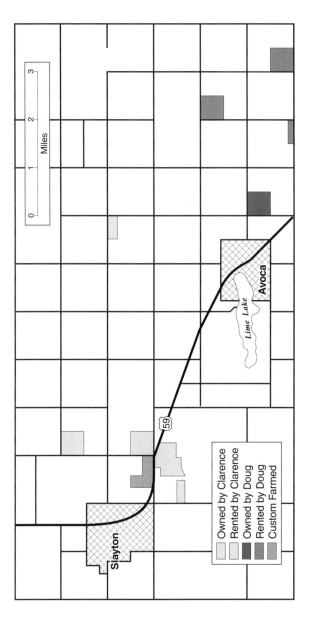

Fig. 12.9. In 1982 Doug and Brenda Magnus farmed 560 acres in southwestern Minnesota. They lived on a 160-acre farm they rented from his father, Clarence; they were buying a 160-acre farm; and they rented 240 acres. Clarence owned a 280-acre farm and rented 300 more acres, and he and Doug custom-farmed a 160-acre farm. Clarence, Doug, and Brenda did all the work themselves, and hired no labor. Reproduced by permission from John Fraser Hart, *The Land That Feeds Us* (New York: W. W. Norton, 1991), 158.

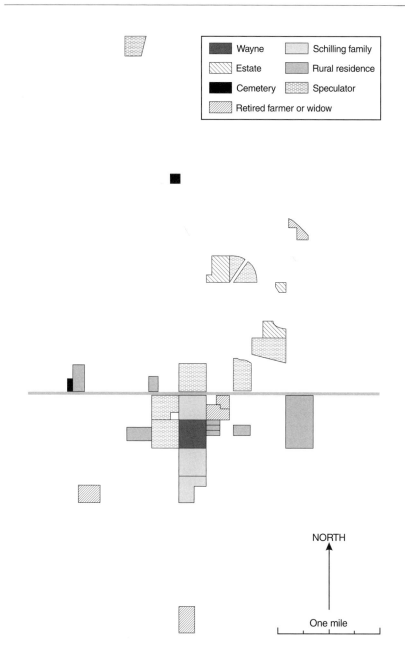

Fig. 12.10.   In 1995 Wayne Schilling was the last remaining dairy farmer in a rapidly growing suburb of the Twin Cities of Minneapolis and St. Paul, Minnesota. He managed to stay in business by putting together a farm of 540 acres by renting twenty-eight separate parcels of land from as many different owners.

Farms will continue to grow larger and the number of farms will continue to shrink. In 1935 the United States had nearly seven million farms, but in 1992 it had fewer than two million, and only about one-third of a million of these "farms" were actually viable, sustainable, self-supporting operations. They can produce all the food and fiber the nation needs, plus a healthy surplus to export.

The most obvious manifestation of farms is their farmsteads, and between 1935 and 1992 roughly five million farmsteads in the United States became redundant in the farm economy. Some have been razed, some have simply been abandoned, and some have been recycled into rural residences. Many redundant farmsteads are deceptive because the old farm buildings still stand, and they still look like farmsteads. The farmsteads on nonfarm farms can be even more deceptive; some of the buildings are still in use, and they look for all the world like real farmsteads. It takes a keen eye to tell them apart.

Part Five

SMALL TOWNS

AND

THE URBAN EDGE

In most parts of the world, farmers live in nucleated agricultural villages, and dispersed farmsteads are unusual. In North America, however, most farmers live on their own land in dispersed or isolated farmsteads, although some farmers do live in agricultural villages in New England, in the Mormon oases, in the Hispanic Southwest, on the Canadian prairies, and in a few other areas.

In North America most small towns and villages were founded with the assumption that they would be collecting, processing, shipping, distribution, and service centers for the agricultural areas around them, and all settlement clusters have some urban function. The replacement of the horse and buggy by the automobile has diminished their retailing function and increased their manufacturing function.

The automobile has also allowed city people to spread far and wide through the countryside in search of residences and recreation. The encroachment of the city on the countryside is like the bow wave of a ship. The bow wave is the standing wave that always remains just in front of the bow of the ship as it moves through the water. A bow wave of intensely cultivated, high-priced agricultural land remains just in front of the expanding built-up area of the city. Some farmsteads that have been rendered redundant by farm enlargement have been abandoned, but many of those near cities have been recycled into nonfarm residences. The spread of city people into the countryside has muddled popular thinking about rural, farm, and nonmetropolitan areas, although conceptually such areas remain distinct.

City people increasingly view rural areas as places for recreation and retirement. Traditionally people of wealth have escaped to the countryside, but the automobile has enabled Everyperson to enjoy it more, and the most rapidly growing rural areas are the metropolitan fringes and popular resort and retirement areas.

13 *Small Towns*

Most settlements in North America were founded with the hope and expectation that they would become centers of commerce, but in the rest of the world, including Europe, most settlements were originally agricultural villages where peasant farmers lived, and commercial functions have developed only gradually in those that have become market centers. Many simple societies do not produce surpluses large enough to support the permanent stores and market centers of highly specialized commercial economies.

MARKETS

Every human group needs marketplaces where its members can exchange their surplus goods. The theory of a subsistence economy, in which people produce all they need and need all they produce, is an interesting conceit of economists, but a true subsistence economy probably has never existed anywhere in the world at any time. A team of sociologists exposed the fallacy of this conceit with unintentional irony when it concluded that contemporary rural communities of native people in southeastern Alaska must have cash income to buy the inputs necessary to support their subsistence economy and lifestyle.

Every family that cultivates the land, even if no more than a backyard garden, at some time during the year will produce more than it cares to consume or preserve. A modern family may try to curry favor with neighbors by giving them vegetables fresh from the garden, but in simpler societies the family might try to trade its surplus for things it needs. Once a week everyone forgathers at the local marketplace, because the weekly market is also a social event, but for the other six days of the week the market-

Fig. 13.1.   The weekly market in Hlabisa, South Africa.

place may be as deserted as a football stadium on a Wednesday morning in February.

Bartering, the simplest technique for exchanging surpluses, is still common in many parts of the world (figs. 13.1 and 13.2), but in modern societies the exchange of money has almost completely replaced the physical exchange of goods. Even in the United States, however, many a boy who grew up on a farm can remember going to the crossroads store with his grandmother and watching her trade a basket of eggs for goods that the family could not produce on the farm.

The best place to make a good swap is where the largest number of potential swappers are gathered together, and in medieval Europe that place was the churchyard on Sunday morning (fig. 13.3). The presence of the church might also have given the swappers a feeling, perhaps unwarranted, of greater safety from sharp dealers and thieves. In other parts of the world the congregation of the faithful, plus an aura of sanctity that helped maintain the peace necessary for orderly business transactions, has conferred special advantages on holy places, such as churches and shrines much visited by pilgrims, as sites for markets and fairs.

The association between religion and commerce has not always been

Fig. 13.2. The weekly market in Brecon, South Wales.

a source of comfort and satisfaction for the church, however, because markets can become rowdy affairs. "There is a widespread belief, by no means confined to Ireland, that fair days are always rainy days," said Estyn Evans in *Irish Folkways*, "as a consequence of the lies and profanity, the fights and pagan ways of the fair," and certainly the traditional activities at Donnybrook Fair, to cite an extreme example, were hardly conducive to quiet meditation and devotion. Modern blue laws, which are motivated by a pious desire to maintain the dignity and propriety of the Sabbath against commercial intrusion, have antecedents more than a thousand years old. Various English monarchs, beginning as early as A.D. 920, have forbidden the holding of markets on Sunday, and in 1285 Edward I even felt compelled to issue an edict against holding them in churchyards, but it is seldom safe to regard medieval legislation as an index to anything more than the intentions of the legislator.

The casual gathering beside the church remained inchoate in many villages, but in some it became more formalized, with the same set market day each week. Itinerant merchants began to make it one of the regular stops on their circuits, and in time the village burgeoned into a market town. When the market prospered it came to the attention of the medieval

Fig. 13.3. Many market towns in western Europe grew up around churches. The distant church dominates this small market town in southwestern France.

rulers, who were always strapped for cash, and they realized that it could become a source of income. They required the market town to pay a tax for the privilege of holding the market, but in return the town received a charter granting it the right to monopolize the trade of the surrounding area.

The holders of market charters, desirous of preventing encroachment on their monopolies, secured legal agreement to the principle that no other charter would be granted within a minimal distance of six and two-thirds miles, or one-third of a reasonable day's journey of twenty miles. This permitted even the most distant person to reach the market in the first third of the day, spend the middle third doing business, and still have time to return home before nightfall. Despite this legal restriction, however, many markets in the richer parts of East Anglia were less than four miles apart, and the same was probably true in other prosperous parts of England.

In England the earliest markets were merely large open spaces where merchants might erect temporary booths or stalls, but in time these were replaced by more-permanent structures, and the modern marketplace is lined with stores. The open-air market has not yet disappeared, however, either in England or in other parts of Europe, and traders still set up their temporary stalls in the open marketplace on a regular market day each week, exactly in the medieval fashion. Open-air markets also continue to thrive in the larger cities of Europe, as well as in the smaller country towns; in Paris, for example, the green vegetables you buy in the open-air markets are as fresh as, if not fresher than, those on sale in the greengrocery, and it is much more fun to shop in the market.

One of the distinctive structures of many marketplaces is the market cross, at once a reminder of the church under whose protective shadow the market was founded, an emblem of the authority by which the charter was granted, and a symbol of "the peace of the market," whose enforcement was required by the terms of the charter. The market cross might be no more than a simple stone spire rising from a low dais, which provided a convenient platform for the reading of proclamations and the like, but some were elaborate vaulted structures with an open space beneath where people could take shelter from the weather; in Chichester, for example, the large and ornate market cross at the intersection of the two principal streets is an impressive structure, but it is a positive menace to traffic in the age of the automobile (fig. 13.4). The best-known market cross, which is associated with the nursery rhyme about seeing "a fine lady upon a white horse," is in the Oxfordshire village of Banbury. Banbury Cross stands at the end of a large open space that is still called the Horsefair, even though it is now a public car park (fig. 13.5).

FAIRS

A market was commonly held on the same day each week, and it attracted mainly local people. A fair was held less frequently, perhaps once a quarter or once a year, but it lasted for several days and attracted traders from greater distances. Products that ripened or became available only at certain seasons, such as wool or wheat, were sold at the end of the season in a great fair, whereas routine commodities, especially perishables, were exchanged at the regular weekly market. Many fairs, like many markets, originated in association with religious observances, and often a fair was held during the festive period dedicated to a particular saint.

One of the responsibilities of wise and holy men in preliterate society was keeping the calendar, and the festivals of specific saints served the practical purpose of reminding people about important dates in the annual cycle of agricultural production. In Ireland, for example, Saint Patrick's Day

Fig. 13.4.   An elaborate market cross in Chichester, England.

Fig. 13.5.   The market cross in Banbury, England, is famous because of the nursery rhyme.

marks the end of winter and the time to start the year's work in the fields, and All HallowE'en (Halloween), the evening preceding All Saints' Day, was the end of the agricultural year, after which ghosts and goblins were free to roam the fields again. An end-of-harvest festival obviously was a good time for a fair because it assembled both the people and the fruits of their harvest.

Andre Allix, in "The Geography of Fairs," distinguished livestock fairs, in which local country people still get together at regular intervals to trade their surplus animals (fig. 13.6), from the great medieval commodity fairs, which were held at neutral frontier sites on the routes connecting different areas, and which laid the foundation for the first commercial cities. The commodity fairs provided the sole mechanism for large-scale interregional exchange when long-range transportation was both difficult and dangerous. They flourished in Flanders and Champagne between the twelfth and the fourteenth centuries, and then gradually migrated eastward as transportation facilities in western Europe were improved and made more secure. The American agricultural fair is similar to European fairs in name only, because adequate facilities for exchange already existed at the time when it was founded, and its principal objective has been to educate local farmers in the latest scientific advances in agriculture.

CENTRAL PLACES

In North America most settlements, from the largest metropolis to the smallest hamlet, were originally founded to serve as collecting, shipping, processing, distributing, and service centers for the surrounding countryside, and most small towns, villages, and hamlets still perform some of these functions. Their collecting and shipping facilities include elevators in grain-growing areas, sales warehouses in tobacco areas, stockyards in livestock areas, and woodyards in forest-products areas.

The small town was an essential link between the rural area it served and the world outside. The first small towns were on transportation routes, first waterways, later railroads, for easy access to distant places (fig. 13.7). A bridging point, where overland routes converged to cross a navigable waterway, was a favored location in the early days because it had good access to two transportation media.

Some of the products of the land must be processed to reduce the cost of shipping them by reducing their bulk, and small towns have creameries and cheese factories in dairy areas, gins in cotton country, peanut-drying

Fig. 13.6. A bull fair in Chateauneuf du Faou, France.

plants, flour mills, sugar mills, sawmills, canneries, wineries, smelters, and refineries. Waterfalls or rapids that could be harnessed to power such establishments were especially attractive sites for small towns in the early days (fig. 13.8), and places at unusually good power sites grew into important manufacturing centers.

Small towns are also distribution and service centers where people can obtain the goods and services they cannot produce for themselves. The traditional heart of the small town is the stores clustered along Main Street, which became the symbol of small-town rural America. Above the stores are the professional offices of doctors, lawyers, and dentists. Main Street has many other service establishments, including banks, hotels, movie theaters, barbershops and beauty shops, the post office, and a variety of eating and drinking places. Nearby are schools, churches, the library, the city hall, the county courthouse, the hospital, and a newspaper office and print shop.

Some of these establishments are basic, so called because they bring money into the place and help support it, but equally important are the nonbasic, or housekeeping, establishments that are essential for the efficient operation and safety of the place. Nonbasic activities include police and fire protection, water supply, and sewage systems. In practice it often is difficult to determine the degree to which a particular establishment is basic or nonbasic, because many are a bit of both, but the theoretical distinction is useful in identifying what makes a place prosper.

Every establishment has a threshold population, which is the number of customers it needs in order to stay in business. Low-order establishments, which have low threshold populations, provide convenience goods and services that are needed frequently, perhaps even daily; they are numerous and spaced fairly close together because their customers need to obtain their goods and services with a minimum amount of travel. Higher-order establishments, which have higher threshold populations, provide shopping goods and services. They are less numerous and are spaced farther apart because their big-ticket items are bought less frequently and justify longer trips and more serious thought before they are purchased.

A natural hierarchy of places, ranging from hamlets through villages and towns to cities and metropolises, is based on the range of goods and services they offer. Larger places have more people, more establishments,

Fig. 13.7. Many small towns in the western Midwest started as distribution centers on railroad lines. On the prairies the lumberyard and coal dealer provided the building material and fuel that farmers farther east had been able to get from their own farm woodlots.

Fig. 13.8. Waterfalls provided power that enabled a small town to become a processing center.

higher-order establishments, and a greater range of goods and services than smaller places. Small places do not have enough customers to support higher-order establishments, and one of their serious disadvantages is their lack of enough people to support the full complement of desirable goods and services that are available in larger places.

It is exceedingly difficult to generalize about the specific ensemble of goods and services that one might expect to find in a place of any given size, because establishments are not restricted to places of a specified minimal size. Furthermore, these establishments reflect the needs and the wishes of the people of a given region at a given time. For example, one does not expect to find grain elevators in Maine, cotton gins in Montana, or creameries in Mississippi; taverns are common in the smallest central places in Wisconsin, where drinking beer is one of the favorite indoor pastimes, but they are rare in southern and eastern Kentucky, where the majority of the people believe that beer drinking is sinful, and the sale of beer is prohibited by law.

Nevertheless, it is possible to make a few generalizations about the size of the population and the functions associated with the smallest central places in the American Midwest. Hamlets, at the bottom of the hierarchy, have a population of some one hundred people. A hamlet commonly has a grocery or general store, a filling station, a café or tavern, an elementary school, a community church, and perhaps a post office and combination feed store/lumberyard/coal yard. Small villages, one step above hamlets, have a population of approximately five hundred people. A small village provides the same goods and services as a hamlet, and in addition it has a hardware/appliance store, a barber, a beauty shop, an implement dealer, a garage, and perhaps a drugstore and an undertaker. A large village, with a population of around a thousand people, adds a furniture store, a clothing store, an automobile dealer, a doctor, a dentist, a bank, a high school, and perhaps a hotel or motel, a movie house, and a weekly newspaper. It does not have a florist, a liquor store, a jewelry store, a department store, a public library, or a hospital.

Every establishment and every place has a trade area from which it draws its customers. The trade area of a place is the composite of the trade areas of all its establishments. One might think that trade areas would be more or less circular, but in fact circles are inefficient space-filling geometric forms because they leave gaps if their outer margins barely touch and they overlap too much if they completely cover an area. Hexagons are much more efficient, and in theory, at least, trade areas ought to be hexagonal.

Theory predicts that a flat, featureless, homogeneous area would be mantled with a network of hexagonal trade areas, with a place at the center of each hexagon (fig. 13.9). A place of the next highest order would occupy the center of each hexagon of low-order places, and so on up the urban hierarchy, with the entire urban network dominated by a place of the highest order. The world is not flat, featureless, and homogeneous, so this theoretical pattern is greatly modified by mountains and rivers, railroads and highways, state and county lines, and many other factors.

POPULATION GROWTH

Few contemporary Americans have ever been inside a real blacksmith shop, and even fewer would know that a livery stable was an establishment where you could rent horses and vehicles, but at the turn of the century every place had at least one of each. The replacement of the horse by the internal combustion engine hurt small towns, villages, and hamlets in at least three major ways. First, some former activities became obsolete and

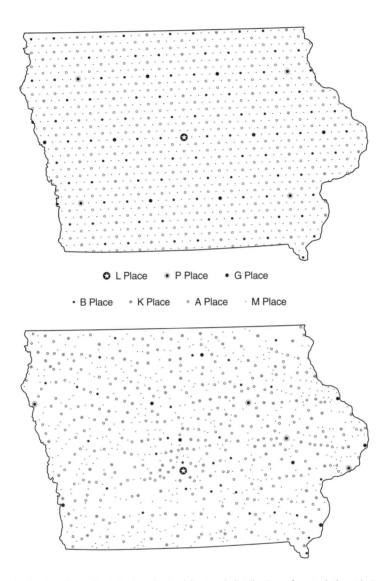

Fig. 13.9. The theoretical (top) and actual (bottom) distribution of central places in Iowa. The theoretical map has a network of nested hexagons; a place of the next highest order occupies the center of each hexagon of lower-order places, and the highest-order place dominates the entire urban system. The actual distribution of central places in Iowa is strongly influenced by railroad lines. The map of the theoretical distribution is reproduced by permission from W. A. V. Clark, G. Rushton, and R. G. Golledge, "The Spatial Structure of the Iowa Urban Network," *Geographical Analysis* 2 (1970): 302. Copyright 1970 by the Ohio State University Press. All rights reserved. The map of the actual distribution is reproduced by permission from *Geographical Review* 78 (1988): 273.

simply disappeared. Second, tractors enabled farmers to handle larger acreages of land, thus increasing the size of farms and reducing the number of workers needed on the land and the number of potential farm customers. Third, the automobile for shopping and the truck for hauling goods to market have increased the range and mobility of rural people.

The smallest central places originally were spaced at horse-and-buggy distances, and many became redundant when cars and trucks enabled rural people to travel greater distances to get the goods and services they needed (fig. 13.10). Many of the retail and service activities of small places gradually shifted up the urban hierarchy to larger places, and out to the malls and shopping strips that festoon them. Today Main Street, in places large as well as small, has become little more than a convenience shopping area for the people who work there.

The long, slow, lingering death of Main Street was well under way before World War II, and today Main Street is a depressing sight (fig. 13.11). The bank on the corner has become a beer joint, and weeds are popping up through the cracks in the sidewalk. Some stores have been boarded up, and others have been converted into private residences with curtains across their windows. The handful still open for business obviously are struggling.

Theorists have assumed that the small central places serving agricultural areas must have been losing population, because they are economically moribund. If the economic function of a place is related to the size of its population, it seems logical to assume that the population should decline when the economic function withers away. No matter how logical this assumption might be in theory, however, it simply has not been supported by the facts, because most small towns in the Midwest have continued to gain population, albeit slowly and fitfully, during most of the twentieth century.

The growth of the population of a random sample of four hundred small incorporated places in the Midwest shows that the erratic and unpredictable growth of individual places has disguised an impressive degree of stability in the entire system (fig. 13.12). The population of any given place has fluttered up and down, up at one census, down at the next, but it has remained within a fairly narrow range, and it has kept its relative position in the system. Most places of about the same size were incorporated at about the same time and have grown at about the same rate.

The size of a place at any given census is related to the date at which it was founded and incorporated. The larger places were incorporated during the initial phase of settlement, and they have continued to grow. Their rate

Fig. 13.10. The Model T Ford forced the village blacksmith shop to become a garage, and it inaugurated the shift of retail and service activities up the urban hierarchy to larger places.

of growth has been appreciably slower than the national rate, however, and cynics might describe it as "upward stagnation" rather than growth, but the fact remains that these places have continued to gain population.

Smaller places later filled in the gaps between the larger places, and most of them also had been incorporated by the turn of the century. They too seemed doomed to continue growing if they reached a certain minimal population, which was somewhere between 250 and 500 people. The only places that have consistently lost population either are in mining areas or were founded so late that they never managed to grow to the minimal population necessary for continued growth.

It is easy to identify a place that has attained this minimal population because it almost always has a solid block of contiguous buildings, usually two-story structures of brick or masonry, on at least one side of Main Street (fig. 13.13). In smaller places even the principal business street has a gap-toothed look, with open spaces and grassy plots between the individual buildings (fig. 13.14).

In the state of Minnesota the population of the entire universe of free-standing incorporated places shows the same overall pattern of slow growth when places are grouped by size (fig. 13.15), but the geographical pattern

Fig. 13.11. Main Street has been dying a long, slow, lingering death ever since the automobile enabled people to shop elsewhere.

of change has been complex (fig. 13.16). The population of places in northeastern Minnesota has fluctuated with the fortunes of the mining and forest industries, and the railroad towns of the southwestern prairies were too late, too many, and too small; few ever reached the minimal population of 350 people that seems necessary for continued growth in Minnesota. The central part of the state, however, had a mixture of places that gained or lost more than 5 percent, and more places gained than lost.

Some people, when told that many small places had continued to gain population in the 1970s, immediately jumped to the conclusion that the growing places had to be dormitory communities from which residents commuted to the rapidly growing fringes of major cities, but many of these places were far beyond commuting distance, and commuting does not explain their growth nearly so well as does their mere existence. The size of a place seemed to predict the growth of its population.

The optimism of the 1970s had to be tempered when the results of the 1990 census became available, because the 1980s were the toughest decade in history for small places in Minnesota. Half of them lost more than 5 percent of their population, only a quarter gained, and three-quarters of those that gained were within commuting distance of major cities (fig. 13.17).

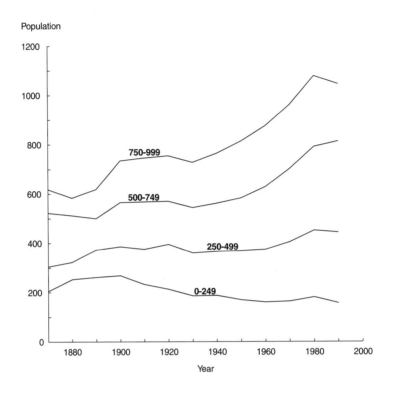

Fig. 13.12.   When a sample of small places in the Midwest is grouped by the size of their population in 1960 (or in any other year), they seem to be stagnating upward.

Geography apparently has replaced size of place as the best predictor of small-town population growth in Minnesota, and Everett G. Smith Jr. has found the same change in Illinois. The long-awaited demise of small towns and villages beyond the metropolitan periphery, which has been confidently predicted for half a century or so (while they have actually continued to gain population) may finally be at hand.

WHY?

Social Capital

Why did small places continue to gain population for so long, despite the confident predictions that they were losing? First, it is important to remember that people live (as well as make a living) in these places, and they do not tear down their houses when stores close on Main Street. The social capital—hopes and dreams as well as hard cash—invested in the homes, schools, churches, and other structures of a place is a major conservative factor in maintaining its population even when it is losing economic activities. Although they may lack some modern conveniences, many village houses are large, and they are astonishingly cheap (fig. 13.18). Cheap housing has attracted low-income workers with large families, who can commute to distant jobs, and the residentially footloose, who are not tied to a specific place of residence: the retired farmer, the person living on Social Security, the divorcée trying to raise a large family on skimpy alimony or welfare payments, the salesman who may live anywhere in his territory. Perhaps, you may think, there are not a great many such people, but it does not require many to maintain the population of a village of fewer than a thousand people.

The Dispersed City

The dispersed-city hypothesis helps explain the persistence of many small places by postulating that they are less moribund economically than

Fig. 13.13. A place that has attained the minimum population necessary for continued growth usually has a solid block of contiguous buildings on Main Street.

Fig. 13.14. Main Street looks gap-toothed, with empty spaces between the buildings, in places that have not yet attained the minimum population necessary for continued growth.

they might seem. Quantitatively oriented students have been preoccupied with the number of establishments and functions in small central places, and they have paid too little attention to their quality, which is vastly more important in a small place with few businesses than in a larger place with many.

The number of functions in many small places undoubtedly has declined since horse-and-buggy days, and these places lack the total mix of functions requisite to a complete rural service center (if they had them all they probably would be bigger places!), but at least one function has survived and flourished, and often it has flourished quite handsomely. Many small places have outsized functions, activities that are far too large to be supported by the number of customers in their immediate trade areas. They draw customers from much larger areas than the trade areas of the places themselves.

Many small places are dominated by a single outsized function, and their residents drive to other places for the other goods and services they need. For example, a man who lives in village A and works in a factory in village B will think little, on a nice spring Saturday afternoon, of driving to the new supermarket in village C to buy groceries; to the area's best hardware store, in village D, to look over a new power saw; to the automobile dealer in village E to find out whether he can wangle a better deal than he could make in Big City; and then stop off, on the way home, at the tavern in village F to have a beer with the boys in the best saloon around. It beats sitting home alone watching a lousy baseball game on television.

In other words, traditional farm trading centers seem to be sorting themselves out into specialized centers dominated by one activity, or only a few, but there is no reason this should be considered unusual. A few generations ago most city dwellers lived close to, if not actually in, the structures in which they did their work, but the two functions of residence and work became separated as cities grew in size and complexity, and as transportation improved. The various functional areas of a modern city are highly localized, and often they are separated from each other by some little distance. Private homes are uncommon in shopping centers, factories are rare in the central business district, department stores are not in industrial districts, and neither stores nor factories are welcomed as neighbors

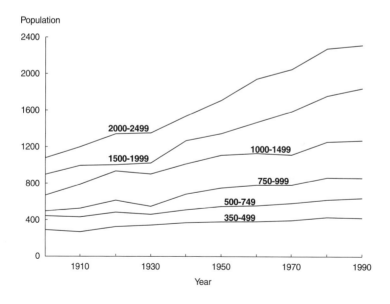

Fig. 13.15. When all incorporated places in Minnesota are grouped by the size of their population in 1980, the entire urban system of the state seems to have been growing slowly. Reproduced by permission from *CURA Reporter* 21, no. 4 (1991): 9.

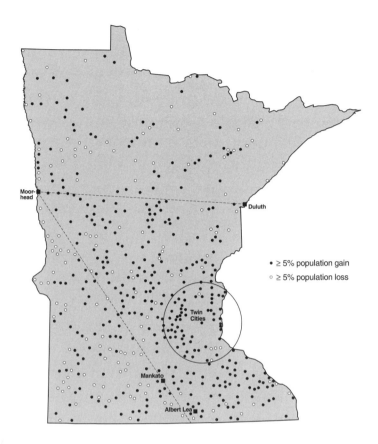

Fig. 13.16. The population change of incorporated places in Minnesota between 1970 and 1980. Reproduced by permission from *CURA Reporter* 19, no. 3 (1989): 2.

in the better residential areas. Each functional area serves the entire city, or some large portion of it, and not just those people who happen to live nearby. City dwellers expect to live, work, and shop in different places, and they spend a fair portion of their time traveling from one functional area to another. Why should anyone expect dwellers in the country to be different?

Manufacturing

Many small places have continued to gain population because their basic reason for being has changed; manufacturing has become ubiquitous, and is the best example of an outsized function. Since World War II, employment in manufacturing has been quietly trickling down the urban hierarchy while retail and service activities have been shifting upward, and many small places have been transformed almost imperceptibly from central places serving their surrounding agricultural areas into small cogs in the national system of manufacturing centers. This change has been well concealed in aggregate employment data for the nation and even for individual states, however, because these data are so heavily weighted by metropolitan areas that have been losing jobs faster than small places have been gaining them.

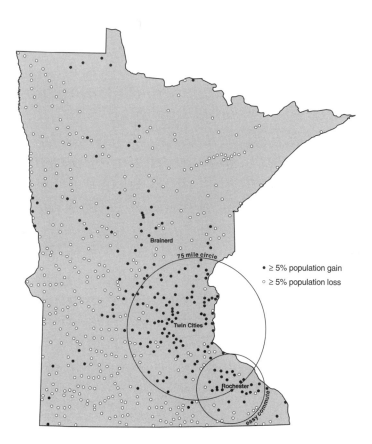

Fig. 13.17. The population change of incorporated places in Minnesota between 1980 and 1990, when geography replaced size of place as the best predictor of growth. Reproduced by permission from *CURA Reporter* 21, no. 4 (1991): 10.

Fig. 13.18. Many small towns have fine old houses that suffer from deferred maintenance. They can be bought for a song by any do-it-yourselfer who is willing to fix them up.

The new manufacturing plants have been attracted to small places because the construction of the Interstate Highway System and improvements in telecommunications have made them more accessible. In the city, factories have to compete for a shrinking pool of increasingly expensive labor, but in smaller places they can have their pick of a labor force with a highly developed work ethic, and they can pay lower wages. Taxes and land prices also are lower in small places.

Some of the new manufacturing establishments are branch plants of large corporations, but an impressive number are homegrown, the brainchildren of local entrepreneurs. Some process local agricultural commodities; many, however, depend on the local area only for cheap labor, and they produce a truly remarkable variety of products (fig. 13.19). Visitors often underestimate the importance of manufacturing in a small town when they see that the planned industrial park at the edge of town is occupied only by lush stands of weeds, because many of the new factories occupy cheap floor space in recycled older buildings, such as former schoolhouses that had to be sturdy enough to keep the students from trashing them (see fig. 1.8), and they are well-nigh invisible to outsiders.

Leadership

Industrial investment slowed down during the recession of the early 1980s, and small towns had to work hard to attract new plants. Some of those that succeeded were startled to discover that they no longer had an adequate supply of labor, because the demographics of rural areas are changing. The number of farms has dwindled, and there is no longer a surplus farm population from which to draw. The remaining rural population is aging; few young people are returning after they go away to college. One of the tough questions that small towns must ask themselves is "Why should any college graduate want to move and live here?"

The birthrate in rural areas is declining, and their people are not reproducing themselves. The population of rural areas will continue to decline unless people change their breeding habits, which seems unlikely. If population growth is desirable (and some people would argue that it is not), small towns are going to have to recruit immigrants from areas outside the United States, and they will be competing with intervening opportunities

Fig. 13.19.   For a time the former grocery store in Danvers, Minnesota, served as a factory in which local women sewed special girdles for people who had undergone liposuction surgery.

in coastal areas and in big cities that are perceived to be more attractive. Furthermore, many of these new immigrants will be dark-skinned people who speak little English, and many current residents do not feel particularly comfortable with or hospitable toward them.

The future of small towns depends on the ability of their leaders to adjust to new people and new ideas. Too often both the minds and the ranks of small-town leadership are closed to women, to young people, and to outsiders. The leaders must understand that small towns have changed from rural service centers to minor manufacturing centers, and attempts to pump new life into Main Street probably are doomed.

Small towns must learn to cooperate instead of competing, which will not be easy after decades of intense rivalry on the football field, the basketball court, the baseball diamond. Just as small places can form a dispersed city in a rural area, so larger places can form a dispersed metropolis. No one place can hope to have everything, but one can claim the regional hospital, another the regional airport, a third the regional library, and so forth.

A small town is what its people make it. It has grown over the years because of the dedication and commitment of its individual citizens, their belief in the town, and their willingness to invest their time, energy, money, hopes, and dreams in helping make it prosper. The decision to incorporate, which is closely related to the present size of the town, was a major act of civic commitment. The construction of a solid block of buildings on Main Street, the hallmark of places that have grown, was a major commitment by individual citizen investors.

The next major commitment of the small town must be to do whatever is necessary to draw new industry, to encourage local entrepreneurs, and to make the town attractive to a new and different kind of people. Small towns need new industry, but industry needs small towns almost as much, and it will be attracted to those that understand and satisfy its needs.

14   *The Long Shadow of the City*

The United States and Canada are heavily urbanized countries. Two of every three Americans live in an urbanized area, and one of every four lives in one of the ten largest. Four of every ten Canadians live in one of the six largest urbanized areas in Canada. No part of the United States and precious few parts of Canada have managed to escape the demands of these city people.

Those who stick to the major highways and the popular media believe that a tidal wave of development is sweeping inexorably outward from our cities to inundate the countryside—that not much open space is left. They routinely castigate uncontrolled urban expansion because they say that providing necessary services to scattered areas is too expensive. They complain that urban expansion deprives the nation of irreplaceable farmland, and that it is aesthetically offensive.

The physical impact of urban expansion is magnified psychologically in that new developments must be accessible and highly visible in order to be successful, and they are concentrated most heavily near those cities and along those highways where the greatest numbers of people can see them. The largest cities are expanding most rapidly, and they have the greatest numbers of people to observe this expansion. It is easy to project the experience of New York City or Los Angeles to the entire nation, and difficult to get people to reject the evidence of their own eyes or to convince them that their own experiences and observations are not an adequate basis for generalization.

URBAN ENCROACHMENT ON RURAL AREAS

As a general rule, the impact of the city decreases with distance, although this rule has almost as many exceptions as it does manifestations.

Cities have their greatest impact on rural areas in the rural-urban fringe at the edge of the built-up area of the city, where some of the most intensive agricultural uses of land are being converted to more-lucrative urban uses. The conversion of agricultural land to nonagricultural uses is the result of a natural economic process that is almost unstoppable because growing cities need more land and urban users can afford to pay more for land than can rural users.

The urban edge is the newest frontier of settlement. It is wild and chaotic, and it is changing feverishly, virtually overnight. Like the traditional frontier, which was ruled by raw power, the power of the knife and the gun, this newest frontier also is ruled by raw power: the power of the purse, the almighty dollar. The sudden avalanche of people and money has overwhelmed and paralyzed the traditional political and legal institutions of rural society.

The newest frontier is a place where fortunes are made and lost. It entices those who are willing to put large amounts of money at risk in the hope of making far more. Some people denigrate these risktakers as speculators and developers, but with equal fairness they may be described as entrepreneurs, and society would be the poorer without them.

The raw vigor of the frontier has always offended the sensibilities of effete folk from older, more settled, more civilized areas, and the newest frontier is no exception. Its critics find only noisome chaos. They fulminate against its casual, unplanned, haphazard character. They complain that it has been shaped by a host of individual, pragmatic, economic decisions rather than by any overarching aesthetic vision, that it is the product of raw commercial forces, not of any ideal.

The critics are wrong. The rural-urban fringe actually is far more orderly than most people realize, but its order arises from the wishes and desires of ordinary, individual human beings, as manifest by the hidden hand of the market, and not from their institutions nor from the wisdom of those who consider themselves sages. Some people will demur that the rural-urban fringe is shaped by the decisions of developers, but developers are not free to make capricious decisions. Their primary goal is to make money, lots of it, and they can make lots of money only if they can produce what their fellow citizens want and are willing and able to pay for.

People want space and they want mobility. They want their own houses on their own private lots, and they are willing to commute long distances in order to have them. They want their own set of wheels and the ability to travel when and where they wish. They like the shopping mall and the bypass strip, but they are reluctant to go downtown because it is too tough to park there. They are willing to put their money where their wishes are, and successful developers are clever enough to give them what they want, and to take their money.

The ceaselessly expanding city dismays people who like high densities and dislike cars. They think that rural is good, urban is bad, and the conversion of agricultural land to nonagricultural uses is especially bad. They would like to freeze the urban edge right where it is, but the only way they can counter the economic muscle of developers is by mobilizing the citizenry and exercising their political muscle to control and forestall development. One of the principal functions of government is to protect the body politic against the worst excesses of the hidden hand of the market. Critics of the conversion of farmland to nonfarm use in the United States have not yet been able to come up with an intellectually defensible and politically persuasive argument that can convince the citizenry at large of the necessity of preserving farmland. Farmland preservation serves a good and useful purpose by retaining open space near built-up areas. The time may come when we will be willing to pay farmers to be museum curators, to serve as custodians and caretakers of an open countryside in which city folk can enjoy pleasant outings, but the amenity argument, although honest, is too elitist to generate much popular support.

Farmland preservationists first used horror stories about the loss of food production, but the United States has the luxury of too much good farmland, rather than too little. The basic problem that has always bedeviled American agriculture has been overproduction. American farmers already can produce more than they can sell, swap, or even give away, and they rarely have to be urged to produce more of anything. Furthermore, the United States is not losing significant amounts of farmland. At current rates of conversion the nation will not run out of farmland for five hundred years. It has been five hundred years since Columbus first set foot in the New World. An awful lot can happen in half a millennium, and it is too early to hit the panic button yet.

Critics complained that they had been misunderstood when their loss-of-farmland argument failed to wash, and they tried the prime-farmland ploy. Perhaps we are not losing large acreages of farmland to urban encroachment, they said, but we are losing our best and most valuable acreages. Again they were wrong. The most rapidly growing counties in the United States actually have less prime farmland than the national average, and they have converted a smaller percentage of it to urban uses, while the areas with the greatest amounts of prime farmland are feeling little urban pressure.

THE BOW WAVE

The critics failed to understand the difference between the price of farmland and its inherent productivity. Intensively cultivated farmland at the urban edge has always had the highest price tag, regardless of its quality, because it is near so many customers. Its value is derived from its location, not from its quality, and its location keeps shifting outward just ahead of the built-up area of the expanding city.

For example, in 1860, 10 percent of all market-garden products sold in the United States were produced within the present limits of New York City, and the area of intensive truck farming and highest-priced farmland has been moving slowly outward ever since. In 1987 the last four acres of farmland remaining in Brooklyn had the astonishing average value of $73,250 an acre; for comparison, the highest county average farmland value in Iowa was $1,251 an acre, and farmers said it was overpriced.

Some people compare the built-up edge of an expanding city to a bulldozer that demolishes all that stands before it, but the analogy of a bow wave is more appropriate. The rural-urban fringe is the bow wave of the built-up area of the city. It is the zone of intensively cultivated, high-priced agricultural land that always remains just in front of the expanding urban edge. The high price and the intensive cultivation of the agricultural land in the bow wave stem entirely from location, not from any inherent quality of the soil. The agricultural activities of the bow wave simply move farther out when the land is converted to urban use, as inevitably it will be (fig. 14.1).

Cities expand fitfully, in different directions at different times and at different rates, so at any given time the rural-urban fringe is complex and discontinuous, but over time it is apparent that farming systems are arranged around a city in roughly concentric belts of decreasing intensiveness. In going outward from the city you pass through a greenhouse belt, a nursery belt, and a vegetable belt into a dairy area.

Greenhouses are so intensive and can pay such high rents that they can compete with the less intensive urban uses of land, and some persist even though they have been surrounded by residential areas. Nurseries need more land, and can pay less for it, but they need to be as close as possible

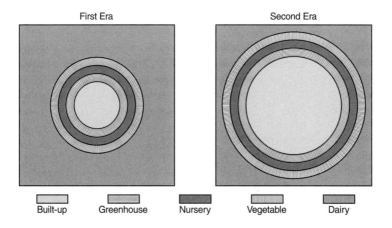

Fig. 14.1.   The bow wave of a city consists of concentric rings of increasingly less intensive agricultural systems. These rings retain their relative position when the expanding built-up area of the city pushes them outward.

to the residences of their customers. They are concentrated near areas of rapid suburban growth and affluent suburbs, where people need or can afford plants to landscape their properties.

The vegetable belt becomes discontinuous as it is pushed farther out, because geometry gives vegetable growers a wider range of environmental options from which to choose; a circle with a two-mile radius has four times the area of a circle with a one-mile radius (see fig. 14.1). Furthermore, since World War II, perimetropolitan vegetable farmers have lost some of the competitive advantage they once enjoyed, because large-scale growers in distant areas with longer growing seasons can deliver truckload lots of uniform produce for a greater part of the year, and vegetable farming is becoming divorced from proximity to consumers.

The supermarket chains that serve the mass metropolitan market demand regular delivery of large quantities of uniform produce for as much of the year as possible, and they would rather deal with one large supplier than with many small local growers. Small-scale vegetable farmers on the rural-urban fringe have had to identify special niches, such as demands for ethnic, organic, and dwarf varieties, and many have turned to direct marketing through roadside stands. Some have started pick-your-own enterprises, but pick-your-own seems to work better with fruits and berries than with vegetables.

Dairy farming, traditionally the dominant type of agriculture in the northeastern United States, constitutes the regional matrix into which peri-metropolitan bow waves have been thrusting. Dairy farmers on the rural-urban fringe are well aware of the pressures and opportunities of the encroaching metropolis, even when it is still beyond the horizon. Individual farmers adopt different strategies in trying to maintain the viability of their farms. Some trade up to more-intensive enterprises, but the anticipation of urbanization deters others from making the necessary investments, and they shift down to activities that are less intensive.

Some dairy farmers sell their land and use the proceeds to buy new farms farther out. Some take off-farm jobs, and shift down from dairy to beef cattle, which demand less time. Some use their farms as bases to develop large cash-crop operations on land they rent from neighbors, from nonfarm landowners, and from speculators who are holding the land for future development. Such leases are insecure, but they can also be surprisingly cheap, because some owners rent for amenity, simply to keep the land looking nice and well tended, as much as for income.

One can humanize the process by thinking about what might happen to four generations on a single farm as the bow wave gradually engulfs it. The first generation continues to milk cows, and grumbles about the encroaching city. At least one member of the family usually has to take an off-farm job, and eventually the farm downshifts from dairy to beef cattle so the entire family can work off-farm. The second generation decides to intensify by growing vegetables, which it sells from a roadside stand. The third generation begins a nursery operation, and may even build a greenhouse. The fourth generation sells the land at a twenty-four-carat price and retires to Florida.

PLUSES AND MINUSES

Farmers in the city's shadow enjoy the advantage of proximity to large numbers of customers, but they have all the problems that other farmers must cope with, plus many that are unique. They have easy access to off-farm jobs, but it is hard for them to hire good labor. They can sell their land to developers at whopping prices, but each departing farmer reduces the number of customers available to support dealers in machinery, fertilizers, feed, and other necessary farm supplies for those who remain.

City people who move to the country create special problems. They drive taxes sky-high when they demand the public services to which they are accustomed. They may not be prepared for the smells, dust, noises, and sprays that are necessary aspects of normal farm operations. One farmer told me that his new neighbors called the police department, the county health department, and the Environmental Protection Agency when he spread manure, and impatient commuters routinely harass his slow-moving farm machinery on narrow country lanes.

Vandalism and litter have driven insurance up in some areas. The digestive tracts of farm animals are not equipped to cope with cans and bottles thrown from passing cars, especially when the cans and bottles get chopped into silage. Teenagers celebrate the dark of the moon by fishtailing their four-wheel-drive vehicles through standing crops, setting fire to fields and buildings, shooting animals, and felling trees. Farmers who try to stop them risk slashed tires and additional malicious damage to machines and equipment, and vandals have even used bolt cutters to make a way through chain-link fences.

REDUNDANT AND RECYCLED FARMSTEADS

The number of farms in the United States dropped from a peak of seven million in 1934 to only two million in 1992, and as many as five million redundant farmsteads may deceive the casual traveler who sees them only from the highway. In rural areas remote from cities many have been razed to reduce the tax bill (fig. 14.2). Near cities, however, many erstwhile farmsteads have been recycled into rural residences, and what looks like a farmstead may be one no longer (fig. 14.3).

In the summer of 1992 I explored the current use of forty-three farmsteads and former farmsteads along fourteen miles of a major highway west of Minneapolis (fig. 14.4). They once served dairy farms of 80 to 120 acres, and each had a traditional dairy barn with stanchions for twelve to sixteen cows and a cylindrical cement silo towering beside it. Two were abandoned, eight still served full-time farms, seven served part-time farms, five were used for nonfarm businesses, and twenty-one were rural residences.

Full-Time Farms

Only three of the eight full-time farms were still dairy farms, and their farmsteads had been greatly enlarged and expanded. For example, Norman

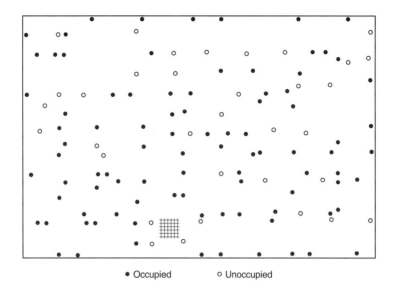

Fig. 14.2. Occupied and abandoned farmsteads in T 102 N, R 54 W, and W 1/2 T 102 N, R 53 W, McCook County, South Dakota, in 1990.

Duske milked sixteen cows by hand when he started farming in 1958, but by 1992 he had extended and remodeled his barn to handle fifty (fig. 14.5). He has built one metal hay shed and shelter on a back lot for his dry cows, and another for steers; he used to sell his steers for veal, but now he feeds them for two years and sells them for beef (fig. 14.6).

"The farm has just gradually kept getting bigger," he said. He has enlarged his original 80-acre farm by buying 160 acres and renting 250 more. He has added two new cement silos and two new metal grain bins to store the additional crops, and he has built two metal machine sheds to shelter the expensive new machines he needs to work the additional land. The dairy barn remains the heart of the farmstead, but like the other two dairy farmsteads, Duske's farm has become an ensemble of new silos, new metal sheds, and new grain bins (fig. 14.7).

Mark Johnson's farmstead looked like a dairy farmstead, but in 1992 he had recently switched to beef cattle. "This eighty acres was my grandma's farm," he said, "and it's so close to the city of Delano that it's only a matter of time until it is going to be turned into housing, so I would be foolish to spend any money on it." He milked thirty-five cows here twice a day for fifteen years, but had to stop when he inherited his father's 330-acre farm because he could not get good help. He grows corn for the beef cattle and cash crops of soybeans and alfalfa. He shook his head when he said to me, "It's amazing what these city people who have moved to the country are willing to pay for quality alfalfa hay for their horses."

It is nice to inherit the land you need to enlarge a dairy farm into a full-time beef-cattle operation, but the more customary strategy is to rent land, as Frenchy Vassar has done. He lives on the fifty-six-acre farm he bought for his father before he died. "You can't afford to own land around here any more," Frenchy said, "so I lease around 600 acres, mainly from other farmers who have quit. My goal is 1,500 acres, but you have to travel too far to find it, and I am already thinking about buying a farm farther from the city."

Frenchy commutes thirty miles each day to his automobile-salvage

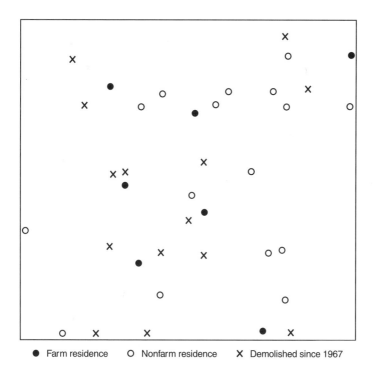

● Farm residence   ○ Nonfarm residence   X Demolished since 1967

Fig. 14.3. In 1967 an area of four square miles in Tipton County, Indiana, had thirty-seven farmsteads. In 1982 seven were still farmsteads, sixteen were nonfarm residences, and fourteen had been demolished. Reproduced by permission from *Geographical Review* 76 (1986): 68.

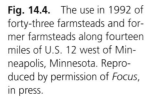

Fig. 14.4. The use in 1992 of forty-three farmsteads and former farmsteads along fourteen miles of U.S. 12 west of Minneapolis, Minnesota. Reproduced by permission of *Focus*, in press.

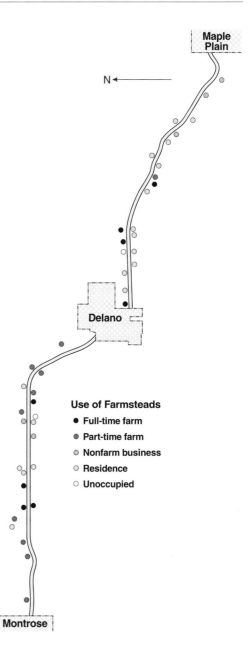

business, but the farm provides full-time employment for Duane Flint, who told me that he is Frenchy's partner. They grow enough corn and alfalfa to finish eighty cattle and four hundred hogs a year, plus soybeans as a cash crop. "Last year we got hit by that big fifteen-inch snowstorm at Halloween," Duane told me, "so we couldn't get into the field to harvest the beans, and I figure that we had to leave more than fifty thousand dollars lying out there in that field."

Flower Farm greenhouse is on a five-acre plot that was once a dairy farmstead. Ned Butterfield bought the farm as an investment. He tore down the old barn and silo, modernized the farmhouse for a rental property, and built the greenhouse. He rents the farmland, eighty acres, to Stanley Duske, Norman's brother, who got tired of being tied down by dairy cows and started a farrow-to-finish hog farm across the road from Norman's farm. Stanley has forty-eight sows, of which he breeds twelve every five weeks, and he expects ten or eleven pigs per litter. "Last week I had 117 born in one day," he said. He owns ninety acres, and uses all the land to grow corn to feed the hogs, but usually manages to have some corn left over to sell.

Al Sterner and Jean Peterson run the Peterson produce farm. They keep their farming as organic as possible, with an idealistic commitment to what they are doing; their operation is the very model of a good back-to-the-land farm. After he finished college Al had various jobs, none fulfilling, and he and Jean got the idea of growing vegetables on her family's 140-acre farm when a budget cut cost Jean her teaching job. Her father had stopped milking cows after he had put his four children through college. He was happy to turn the farm over to Al and Jean.

Al said the operation has gradually gotten bigger. They have a brightly painted roadside stand on the farm, and they sell at nearby stores and farmers' markets. They both obviously enjoy dealing with people. While I was talking to Jean at the stand a car drove up, and she greeted its passengers like long-lost friends. I suspect they bought more than they had intended when they stopped. As they drove off I asked her if they were relatives. She looked at me in surprise and said, "Why, I have never seen them before in my life."

Part-Time Farms

The part-time farmers have off-farm jobs, and do their farm work in the evenings and on weekends. Four of the seven have beef cattle. George Schaust told me that his farm is too close to Minneapolis. "I was milking

Fig. 14.5. Norman Duske converted his sixteen-cow dairy barn into a fifty-cow barn by adding one-story metal sheds to both ends.

thirty-four cows," he said, "and I needed a new barn, but it would have been a foolish investment, because no buyer from the city would be willing to pay anything for a barn." He took a job as the building superintendent at the local school, and has a herd of thirty-five beef cattle that keep him busy on weekends, when the farm is also a mecca for his seven children and their offspring.

In his younger days Joel Zabel worked on his father's dairy farm and also had a full-time off-farm job. "I got up at 3:30 and did the morning milking before I left for work," he said, "and he did the evening milking before I got home. Once when he was sick I had to do both, and it really ran me ragged. He decided to quit farming in 1982, and that's when I switched to beef." Now there are only a couple of bad weeks in the spring and fall when he has to be out in the fields until after midnight and then get up and go to work the next morning, but he expects to retire from either the company or the farm when he turns fifty-five.

Dan Tapio is an idealistic young man and a skilled cabinetmaker. The farm was divided when his stepfather died, and Dan got the twenty-four acres with the old horse barn. He did not want to see the land chopped up into building lots, and he did not have enough land for a dairy farm, but he wanted to be ecologically responsible. He read a lot and considered raising elk, reindeer, even ostriches, until he realized that buffalo would be just right for him.

"A lot of this land around here is in small parcels that are rented out," Dan said, "and they only produce coarse hay. It is not dairy quality, but it is fine for buffaloes." He bought twenty-five buffalo calves, raised them to cows, then bought a bull, and is now producing his own calves. He expects to sell some to other breeders, and will feed the rest for market. The meat is low in cholesterol, he collects and sells the fur the animals molt in summer, and he even bags the manure for sale by a local florist at a handsome price. He keeps the animals where they belong by feeding them well, but he has also built some impressive fences, with woven wire six feet high stapled to old telephone poles.

Nonfarm Businesses and Rural Residences

Five of the former farmsteads are used for nonfarm businesses, including two tree services, an automotive recycling yard, a construction company equipment depot, and a former antique store that the new owner hopes to turn into an amusement center. The remaining twenty-one are

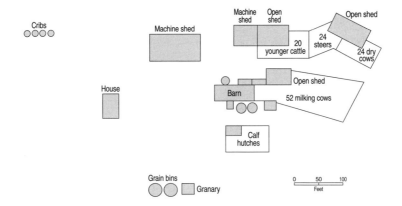

Fig. 14.6. The layout of the Norman Duske farmstead. Reproduced by permission of *Focus*, in press.

Fig. 14.7. The original barn and silo on Norman Duske's farmstead are almost hidden by an ensemble of new grain bins, silos, and metal sheds.

rural residences of commuters. A few travel as much as forty miles each way, but most work in the western suburbs of Minneapolis, and their daily journey to work is no more than ten miles or so.

Most of the farmsteads that have become available to new residents are by-products of farm enlargement. A farmer who has bought a second farm does not need the second farmstead, and usually is only too willing to rent or sell it. Farmsteads on a major highway, despite its heavy traffic, are especially attractive to commuters, and a greater percentage of former farmsteads on secondary roads and in less accessible areas probably remain unoccupied (see figs. 14.2 and 14.3).

I might have expected that the rural residences of city folk would be kept up better than working farmsteads, but a concern for appearance seems to be more closely related to individuals than to occupation. A few former farmsteads are unkempt, but the general level of maintenance is impressive. Dairy farmers in particular like to keep their farmsteads looking as nice as possible, because they rarely get away from them for more than a few hours at a time. Almost everyone seems to have and need a riding power mower. The grass is neatly trimmed, the buildings are freshly painted, and the quality of maintenance is not a criterion for distinguishing working farms from rural residences.

Why have city people moved out to former farmsteads? Each answer is unique, but the two most common elements are elbowroom and the opportunity to keep animals—almost invariably horses, but often dogs as well (fig. 14.8). Judy and Henry Marin bought a fifty-five-acre farm in 1972. "We thought it would be good for our two sons," Judy said, "even though Henry has to work eight hours a day as an electrician just so we can afford to live here. We can have all the dogs and other animals we want, the boys can earn money here, and they don't have to go out and find summer jobs in some fast-food place."

The residents on seven of the twenty-one recycled farmsteads told me that they were "elbowroomers," although that is my word and not theirs. They grew up on the outer metropolitan fringe, and they actively dislike cities and the sense of being hemmed in by buildings and other people. They seem to feel compelled to keep moving farther out when they sense that the built-up area of the city is beginning to crowd in on them too closely. They reminded me of the old saying about the early pioneers who knew it was time to move on when they could see the smoke from another settler's cabin. Elbowroomers have already moved out once, and they are poised to move still farther out if they start to feel too crowded; they are the pioneers of metropolitan encroachment on rural areas.

RURAL RESIDENCE

The massive influx of city people has been the greatest change in rural America during the twentieth century, even greater than the decimation of the farm population. City people need the countryside for residence and they need it for recreation. Before World War I, suburban development consisted largely of the homes of well-to-do people bunched near stops on rapid-transit lines, but thereafter the automobile enabled city people to penetrate virtually every nook and cranny of the countryside. At first they came just to visit, but then they talked farmers into selling them plots of land on which they could build houses, and soon clever developers were buying up and subdividing entire farms.

In 1990 the United States had 61.6 million rural people, but only 3.9 million people lived on farms. The urban side of the coin is even more confusing: in 1990 the total population of the United States was 248.7 million

Fig. 14.8. People move to former farmsteads because they like to have horses and dogs.

people, of whom 192.7 million lived in metropolitan areas, 187.1 million lived in urban places, and 158.3 million lived in urbanized areas. It is clear that the words *rural*, *farm*, and *nonmetropolitan*, which too often are used as though they were interchangeable, actually represent quite different concepts. You cannot use census data intelligently until you understand the official census definitions of the terms it uses.

The census does not even define rural areas and rural people; they are simply the residuals, the leftovers that are not included in the definition of urban. An urban place must have at least twenty-five hundred people; everything else is rural. The threshold of twenty-five hundred was first used in a census report in 1906, and it caught on so well that it has been used ever since, although no one knows why it was chosen in the first place.

A new component had to be added to the urban definition in 1950 because most American cities had sprawled beyond their corporate limits, but the built-up areas outside the city limits still were officially classified as rural. The Census Bureau expanded its definition of urban to include the urban fringe, which is the densely settled (one thousand persons or more per square mile) area contiguous to a city of fifty thousand persons or more. The entire built-up area, including both the city and its urban fringe, is designated the urbanized area (fig. 14.9). The urbanized area keeps expanding as the city grows, and it must be defined anew for each census. It is the best indicator of the actual geographical extent of the city, but it drives statisticians crazy because it grows larger at every census.

A metropolitan area also is based on a city of fifty thousand people or more, although metropolitan status has such advantages that this threshold often is relaxed. Except in New England, a metropolitan area consists of one or more entire counties (see fig. 14.9). Metropolitan areas provide a stable geographical base for data collection and for statistical comparisons over time. It is easy to reconstitute data for metropolitan areas that have been redefined simply by adding or subtracting data for individual counties. Nearly all metropolitan areas include extensive acreages of rural land. Casual users sometimes make the egregious blunder of assuming that *nonmetropolitan* and *rural* are synonymous, but they definitely are not. In 1990 a total of 26.4 million Americans lived in the rural parts of metropolitan areas, and 20.9 million lived in nonmetropolitan urban places. In other words, two of every five rural Americans lived in metropolitan areas, and three of every eight nonmetropolitan people lived in urban places.

As early as 1930 so many city people had moved to rural areas that it

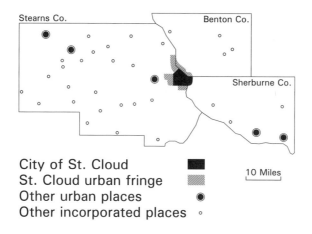

Fig. 14.9. The metropolitan area of St. Cloud, Minnesota, as officially defined by the U.S. Bureau of the Census, includes the central city of St. Cloud, its densely built-up urban fringe, and three entire counties that include five separate urban places and thirty separate incorporated places. It seems clear that the St. Cloud Metropolitan Area includes extensive rural areas, and the true metropolis of St. Cloud includes only the central city and its densely built-up fringe, which are officially defined as the St. Cloud Urbanized Area. Reproduced by permission from Emery N. Castle, ed., *The Changing American Countryside: Rural People and Places* (Lawrence: University Press of Kansas, 1995), 75.

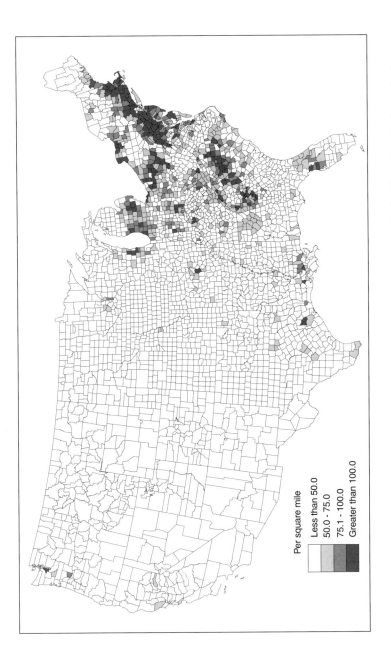

Fig. 14.10. The rural nonfarm population of the United States is officially defined negatively; it includes all people who do not live in urban areas and who do not live on farms. Here, map shows rural nonfarm population in 1990.

Per square mile

Less than 50.0
50.0 - 75.0
75.1 - 100.0
Greater than 100.0

had already become apparent that the Census Bureau needed to distinguish between rural people who lived on farms and those who did not. Rural nonfarm people are a double residual: they do not live in urban places and they do not live on farms. Some of them live in the open country, and some live in places of fewer than twenty-five hundred people, but most of them live just beyond the urbanized areas of places of fifty thousand or more people, or just beyond the city limits of places of twenty-five hundred to fifty thousand people, for which urbanized areas are not defined.

In 1990 the rural nonfarm population of the United States was heavily concentrated near Megalopolis and the other great metropolitan centers of the Northeast (fig. 14.10). It was less dense in the agricultural heartland, where much of the farmland remains in strong hands, and sparse in the West, where the availability of water controls urban sprawl.

The other principal concentration of the rural nonfarm population in 1990 was in the region John Morgan and I have named Spersopolis (the western Carolinas, northern Georgia, and eastern Tennessee), where a new pattern of occupance has developed since World War II. Many rural landowners in Spersopolis have been only too happy to sell off small roadside building plots, and an almost continuous necklace of houses lines virtually every paved road in the region, with expensive mansions in unembarrassed proximity to mobile homes and tumble-down, weather-beaten shacks.

The heterogeneous amalgam of every known variety of single-family housing is unique to the region, but the spontaneous development of this new pattern of dispersed occupancy along highways suggests that it might actually represent the populist ideal of a city that Americans would create if left to their own devices, free to live where they wished and unfettered by planners. It could be the residential equivalent of the bypass strip that is replacing downtown as the ideal commercial area of the contemporary city.

The electric streetcar radically transformed American cities, and the automobile may be doing the same thing. Commuting from distances that might seem extreme is not the chore it appears to be, either in time or in distance, because the contemporary city has been turned inside out and many commuters travel only to the nearest edge, not to the center. Many of the best new jobs are in the office buildings, malls, stores, and factories jostling with the motels, eating places, and filling stations that line the bypass and the major routes at the edge of the city, and they are just as accessible from distant rural areas as from the older, built-up areas of the city.

Carpooling facilitates long-distance commuting from rural areas. The first rural carpools developed informally when several people drove to a convenient crossroads, parked their cars there for the day, and made the long trip to work in a single vehicle. Highway departments in many states have institutionalized this system by creating special carpool parking lots at major intersections, especially at rural interchanges on interstate highways.

In 1990 at least two of every five rural Americans lived within fifty miles of a city of fifty thousand people or more, and a large but unknown number lived within commuting distance of smaller places. The perimetropolitan fringes, where city people are feathering out into rural areas, accounted for a major share of the nation's total population growth in the 1980s, and they seem destined to continue to do so.

15 *Recreation*

City people use the countryside for recreation as well as for residence, and their demands for rural recreation have been increasing in recent years. Most modern Americans work shorter hours for more money than their parents did, with longer vacations than their parents enjoyed. The development of high-speed automobiles and high-speed highways since World War II has enabled them to fan out in ever-larger numbers over ever-widening areas in search of places that are blessed with bodies of water, scenic uplands, and pleasant climates. The demand for outdoor recreation will surely continue to increase in the years ahead.

STATELY HOMES, HORSES, AND FOX HUNTS

Escape to the countryside has been beyond the means of most city folk until recently, but ever since Roman times the idea has attracted those who could afford it. Wealthy Americans have emulated rich and powerful Englishmen, who have been building palatial mansions in rural areas for half a millennium. A gentleman's "stately home" had to have an appropriate setting, so he fenced off a large area around it and converted it into parkland. The park was an open stretch of grass, tastefully studded with scattered clumps of trees and decorated with flocks of cattle or sheep, which added bucolic charm while they mowed the grass.

Parks, like forests, originally were created in England to provide meat for the Norman rulers in an era when domestic cattle were small, emaciated, and practically inedible. Near each of their royal castles the Norman kings designated certain areas "forests," stocked them with deer, and employed professional huntsmen to put venison on the royal table. Great landowners created similar preserves, but they were called "chases" rather than forests. The king owned the deer but not the land, and the royal habit of running deer on other people's land was one of the abuses that led the nobles to demand the Magna Carta. The king's deer were protected by special forest laws, and the most heinous crime committed by Robin Hood and his merry men in Sherwood Forest was the hunting of deer, which was akin to cattle rustling in modern times.

Forest was a legal rather than a botanical term; few forests had much woodland, and few major woodland areas were enforested. Sherwood Forest disappoints many modern visitors, and one pious worthy, after seeing Ettrick Forest on the Scottish border, was moved to comment, "If Judas Iscariot had betrayed Our Lord in Ettrick Forest, he could not have committed suicide for want of a tree from which to hang himself." The vegetation of parks and forests alike has been greatly modified by grazing animals, which browse tree branches as high as the animals can reach, and graze or trample seedlings before they can become established. Animals soon overgraze and destroy the plant species they like best, and encourage the growth of species they dislike.

Although the Norman kings employed professional huntsmen, they did enjoy hunting when they had the time. It is not unreasonable to trace their passion for hunting back to the great Aryan expansion of about 1800 B.C., when nomadic horsemen from Central Asia suddenly began thundering out over much of Eurasia. The Achaeans of Greece, the Hyksos of Egypt, the Hittites of Anatolia, the Aryans of India, the Shang of China all were probably part of this expansion, which has been bedeviled by overtones of racism, because the name *Aryan* came to be applied to the Indo-European family of languages, and then it was assumed that these languages were related to a racial type. In fact, these were not mass invasions by destructive nomads; the number of invaders was so small that they were absorbed into the native population, and the local languages were not changed, as they would have been if there had been mass slaughter and replacement of peoples.

The Aryan invasions, like the Norman conquest of England, were the work of a small number of skillful leaders and organizers, with a few camp followers. Despite their number they had tremendous power, because they had learned to ride horseback, they had invented the wheel and learned to use horses to pull their fighting chariots, and quite probably they had learned the secret of ferrous metallurgy. They could fashion their weapons and tools of iron rather than bronze. They came as conquering horsemen, taking for themselves the best serfs and the richest lands, where they could be supported by a productive but passive peasant population while they indulged in their favorite pastimes: hunting, fighting, and riding horses.

To this day the heritage of the man on horseback, the conquering warrior, remains a remarkably strong force in European ruling life and society. It generated an epic literature of chivalry and a romantic tradition of cavalry, which was ended quite abruptly by the invention of the machine gun. Trousers, which were devised by the horse-riding nomads, were considered the proper garment for the lordly male, and many men objected when women began to wear slacks. Horse racing, which the nomads enjoyed, remains "the king of sports and the sport of kings," although Europeans use adult jockeys whereas the Mongols used small boys as riders. And riding to hounds in pursuit of the elusive fox remains a highly prestigious activity among those who consider themselves aristocrats, or would like to be so considered.

Organized fox hunting originated in England around 1750. Although some apologists maintain that the purpose of fox hunting is to get rid of foxes, which are a nuisance in rural areas, nothing dismays an avid fox hunter more than the prospect of having no foxes to get rid of. In the late eighteenth and early nineteenth centuries many hunting landlords did their best to forestall such a disaster. They created "fox coverts," sheltered breeding places for foxes, by planting spinneys and clumps of gorse in well-chosen spots on their estates. Sending a pack of hounds through a covert was almost certain to raise a fox and guarantee a good day's hunting. Fox coverts are still scattered over the East Midlands; they are often the only trees in sight for thousands of acres, and they are clearly labeled on the one-inch maps of the Ordnance Survey.

Fox hunting was exported to the United States in 1877, when the first pack of hounds was established at Meadow Brook, Long Island. Some hunts in the United States pursue a dragged scent rather than a live fox, but the ritual and regalia are assiduously patterned after the British model: packs of wealthy suburbanites adorn themselves in scarlet jackets, mount their horses, quaff their stirrup cups, and pelt across the countryside in pursuit of a pack of yelping hounds, to the astonishment of innocent passersby. Fox hunting in the United States is a phenomenon of the outer suburban fringes, with its greatest manifestation in the Estate Belt just west of Megalopolis (fig. 15.1).

The Estate Belt is one of the most elegant agricultural regions in the United States. The lush, rolling countryside proclaims wealth and its appur-

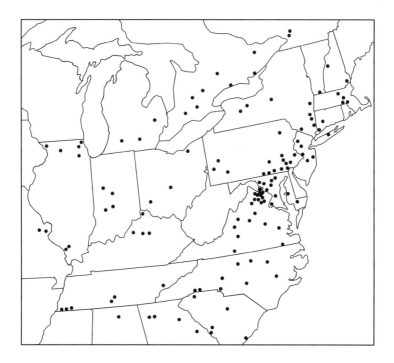

Fig. 15.1. Organized fox-hunting clubs in the winter of 1993–94.

tenances. Registered cattle and purebred horses graze meticulously mani-
cured meadows dappled with shade trees and enclosed by rugged stone
walls or handsome wood fences. The fine old mansions are set well back
from the winding country lanes, discreetly screened by trees and shrubs
from the public gaze. The lands of the estates are interspersed with equally
attractive golf courses, greenhouses, and nurseries, and the neat grounds of
private schools, both boarding and country day. Along the roads are antique
and specialty gift shops, bookstores, and good country inns and restaurants
that cater as much to tourists as to the local residents.

The Estate Belt, which is just west of the world's greatest concentration
of wealth, is merely the epitome of the prestigious suburban area. Every
American city has one, the master bedroom for the city's top brass, where
well-to-do people pay outlandish prices to live on large estates near other
WASP, Ivy League, Republican, fashionable conservatives from some of the
best old families who belong to the best old country club.

The prestige suburbs inevitably attract office buildings and factories, especially the newer kinds with low profiles that can be landscaped to blend unobtrusively into the local scene. The process begins when company presidents suddenly ask themselves why they should have to go in to the office or plant every day when they could just as easily bring it near their homes. Most of the other members of the top management team live close by, and it would be easy enough to pave an adequate parking lot for the workers. In time, the roadsides become festooned with jerry-built ranch houses for the workers, wherever a developer can buy enough land to put up ten or twenty of them.

Financial considerations as well as prestige can draw wealthy people to estate farming, because much of the money they spend on land and livestock can be written off as tax deductions against their nonfarm income. They can be useful members of the agricultural community, however, because they can afford to experiment with new ideas and new techniques, whereas most working farmers cannot. They can also afford to spend money to keep the countryside looking attractive.

The impact of wealth on the American countryside is nowhere more impressive than on the Thoroughbred racehorse breeding farms of the Kentucky Bluegrass area north of Lexington and in the area near Ocala in north central Florida. Both areas have rich soils derived from phosphatic limestone, rolling karst topography, lush, green pastures with white board fences and large, old shade trees, fancy entrance gates, paved farm roads, and handsome horse barns (fig. 15.2). Some owners visit their farms only infrequently, leaving the day-to-day affairs in the hands of skilled resident managers. Both areas have the infrastructure necessary for rearing and training horses, such as tack shops, blacksmiths, special feed stores, veterinarians, a large selection of high-quality stallions standing at stud, and fully equipped training tracks.

The Kentucky Bluegrass country, which is the most beautiful rural area in the United States, is the world's premier racehorse-breeding area. The newer Ocala area has milder winters, which allow the horses to graze outside throughout the year and to train on tracks that are never frozen. Good-quality land near Ocala is not nearly so expensive as land in the Bluegrass, and farm labor is also cheap and abundant. Despite these advantages, however, Ocala was not recognized as a major breeding center until after locally bred horses had won the Kentucky Derby in 1956 and again in 1962, and it will never have the prestige of the Bluegrass.

THE IMPACT OF LEISURE

Before World War II the American countryside belonged to farmers and to wealthy people, but since then other groups have been demanding a share of it. The list of activities in which they engage, in rough order of popularity, includes walking, swimming, bicycle riding, camping, fishing, golf, hiking, running and jogging, hunting, tennis, target shooting, downhill skiing, backpacking, and cross-country skiing. The steadily rising demand has been generated mainly by well-to-do and middle-class people, who are enjoying higher incomes, shorter workweeks, longer vacations, and increased leisure time.

The demand for outdoor recreation varies with age, sex, family size, place of residence, income, education, and occupation. Some people prefer activities that combine mind improvement with peace and quiet, such as nature study, bird-watching, photography, hiking, and sailing. Some people want speed, noise, and blood, and they opt for roaring powerboats, snarling snowmobiles, and the excitement of hunting and fishing. The person roaring past on a snowmobile has contempt for the person on silent

Fig. 15.2. White-painted board fences on a horse farm in the Bluegrass country of Kentucky.

cross-country skis, and vice versa; they are truly people of different cultures, with different value systems.

Outdoor recreation has become big business in the United States. Most estimates are merely educated guesses, but there can be no doubt that opportunities for outdoor recreation have brought hundreds of thousands of people and many millions of dollars into popular resort areas, and recreation has become their economic mainstay. One of the most obvious manifestations of outdoor recreation has been the efflorescence of second homes in resort areas.

The greatest concentration of second homes in the United States is along the Atlantic Coast, from Bar Harbor, Maine, southward to Hilton Head, South Carolina (fig. 15.3). A second major water-oriented concentration is along the scenic rivers and in the lake-speckled morainic areas of the boreal forest in the upper Midwest. Many clusters elsewhere focus on large new bodies of water that have been created by impoundment. Cooler summer climates have attracted people to upland areas in the Northeast, the Blue Ridge Mountains, and the foothills of the Sierra Nevada, and milder winters have attracted them to southern Florida, the Gulf Coast, the lower Rio Grande valley in southern Texas, and the Southwest.

RESORT AREAS

Individual resort areas differ from one another, but some common traits distinguish them from nonresort areas. Americans, when they go on vacation, seem to have four basic needs. They need to sleep, they need to eat, they need to spend money on things they would have sense enough not to buy if they were at home, and they need activities to keep them occupied, because they become restless and uneasy if they are left alone with nothing to do but contemplate nature. Resort areas oblige them with a variety of lodging places, eating and drinking places, souvenir shops, and things to do.

A particular activity or suite of activities, such as those associated with water, might have accounted for the original existence of a resort area, but this attraction has been enhanced by the addition of amusement parks, water slides, golf courses, tennis courts, and many kinds of commercial entertainment. Most resort areas also boast a fine assortment of souvenir shops, antique stores, gift shops, art galleries, and similar establishments— anything that will help tourists get rid of their money.

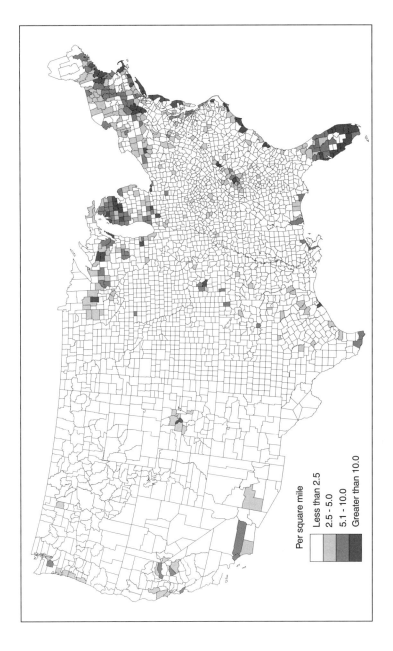

Per square mile

Less than 2.5

2.5 - 5.0

5.1 - 10.0

Greater than 10.0

Fig. 15.3. Second homes in 1990.

Resort areas have many more eating and drinking places than nonresort areas (along with a full line of bakeries, candy stores, fudge shops, and ice-cream parlors) because people do not want to have to bother with preparing meals when they are on vacation. A resort area has an advantage if it can boast an elaborate ceremonial food ritual such as a Cape Cod clambake, a Hawaiian luau, or a fish boil in Door County, Wisconsin, to which all visitors feel obligated to subject themselves at least once during their visit.

Most resort areas have a wide range of accommodations. Some of the earliest resorts were enormous, expensive, elegant grand hotels in attractive settings where people spent their entire vacations. They took a train to the hotel, sat in rocking chairs and rocked for two weeks, and then took a train back home (fig. 15.4). Most modern Americans, even those who can afford them, are too impatient for such leisurely vacations.

At the other extreme from luxury hotels are campgrounds, which are the cheapest places for family vacations. One of every five Americans goes camping each year, and these campers spend an average of ten days or so in their tents, trailers, or camper vans. The big camping season is the school vacation months of June, July, and August. Families seek out campgrounds with woods and water: a shady spot in a scenic area near an unpolluted stream or lake where they can swim, fish, or boat. They probably find one in a national forest or state park. If they stop at a private campground they are likely to demand more: a playground, sports equipment, laundry facilities, perhaps even a heated swimming pool (fig. 15.5).

Many resorts in the north woods originated when farmers let tourists sleep in the haymows of their barns and served them meals in the farmhouse kitchens. Then they built rustic cottages for families, and served meals in a central lodge that was also a social center and game room in bad weather. In order to stay competitive these resorts have had to modernize their cottages, and many have added indoor and outdoor swimming pools, golf courses, tennis courts, riding trails, canoeing, sailing, water-skiing, bingo, movies, bonfires on the beach, and special activities for children.

Some resort areas have become too popular for their own good. Real estate prices and rents have skyrocketed, and many of the people who work in such areas cannot afford to live in them. The service industries generally pay low wages, and those in resort areas are no exception. Summer resort areas often depend on student workers who are willing to work at rock-bottom wages for the privilege of being in a popular resort area, but some employers have to provide dormitories or other accommodations for their student workers.

Fig. 15.4. A traditional resort hotel, with a long front porch on which vacationers could sit in rocking chairs and rock for two weeks.

The ultimate lodging facility is a privately owned cottage. Many summer cottages in the north woods started off as primitive shacks with minimal creature comforts, hunting lodges where men could put on red-checked flannel shirts, play poker, smoke cigars, drink whiskey, hunt and fish every now and then, and pretend they were the reincarnation of Daniel Boone. When women began to come along they almost immediately demanded the comforts of civilization, such as running water, electricity, and telephones.

As the owners aged they realized that the cottages would be great places to retire, so they have winterized them for use as principal residences. Other retired people have established their legal residences in the Sunbelt states, where income taxes are lower, and they routinely migrate between summer homes in the north and winter homes in the south. Some resort-area entrepreneurs also operate winter businesses in the Sunbelt to supplement their summer operations in the north.

The influx of retired people and the maturation of the resort and retirement economies have created better and more-dependable jobs in construction, maintenance (fig. 15.6), and the provision of services, especially

Fig. 15.5. A heated swimming pool makes camping out more tolerable for some vacationers.

health services for older people (figs. 15.7). Local young people no longer have to migrate to the city in search of work when they finish high school, and improved employment opportunities actually have attracted young in-migrants from other areas. Resort and retirement areas have enjoyed above-average rates of population increase since World War II, and they bid fair to remain islands of population growth in the years to come.

Furthermore, the summer resort season, which once lasted only from Memorial Day to Labor Day, has been stretched in both directions, and local people no longer have to earn their entire twelve-month income in four-teen frantic summer weeks (fig. 15.8). Some people like the quieter spring and fall months when children are in school, and many are enticed by a whole new gamut of winter activities that includes snowmobiling, cross-country skiing, and ice fishing. Groups of motels have organized cross-country skiing weeks. You get up in the morning, pack your bags, and set off along well-groomed trails for the next motel, where your bags have already been delivered and are waiting for you in your room.

The extended tourist season has created labor shortages for those resort businesses that used to depend on cheap student workers when the season

Fig. 15.6. The maturation of the resort economy creates jobs in construction and maintenance for local young people.

Fig. 15.7. Many resort areas lack standard municipal services; these are provided by local entrepreneurs.

meshed with the academic calendar. One resort-area newspaper editor bitterly denounced the opening of schools and colleges as an "annual kidnapping," and tourism interests have strongly supported laws that would dictate the fall opening dates of academic institutions.

The landscapes of resort areas can be deceptive because only the local people live where their houses can be seen from the public highways. The whole purpose of a summer cottage is to afford privacy and let the occupants avoid being seen, except perhaps from the water. The cottages have turned their backs to the road, and they are hidden from it by heavy screens of vegetation. Visitors rarely see more than name signs and gravel drives that wind back and disappear among the trees.

In contrast, the commercial establishments in a resort area are as conspicuous as possible (fig. 15.9). Many are advertised by billboards and garish roadside signs because they need the greatest possible visibility in order to attract the casual business on which they depend. Ordinances controlling signs are a constant bone of contention in resort areas, and they are symptomatic of the common disagreements, often acrimonious, about the rate and character of development that should be permitted in a resort area.

At one extreme are the developers who think that anything that brings more money into the area is good. At the other extreme are those who want to pull up the gangplank and keep everyone else out. Many cottage owners want to keep an area precisely as it was when they first saw and were attracted to it, and they resent all who have come after them, often without realizing that they themselves are equally resented by those who have been there even longer.

It is incorrect, however, to assume that length of residence is a dependable predictor of attitudes toward development, because each new issue seems to generate a different alliance. One might expect, for example, that local people would favor development, but a small local business that has been run by members of the same family for two or three generations might be seriously threatened if a large national chain opened a local unit. Conversely, one might expect that owners of second homes would generally oppose development, yet they desire some improvements and better provision of the services they need.

Many second-home owners feel frustrated because they are disenfranchised in one of the two areas where they own homes and pay taxes. A large part of the local tax base in many resort areas belongs to nonresidents and nonvoters who have no voice in local decisions that affect them and their

Fig. 15.8. The development of winter activities has enabled resort areas to extend their seasons.

property. They may not even learn about such decisions until belatedly because they were not resident in the area when the vote was taken.

WISCONSIN RESORT COUNTIES

Despite their similarities, each individual resort area has a distinctive character. Each area identifies and publicizes its own particular image, both to attract the kinds of visitors who will find it congenial and to warn off those who will be disappointed in it, bad-mouth it, and hurt its reputation. Three Wisconsin counties, Vilas, Door, and Adams, show how the attractions and the clientele of resort areas differ. A person who likes any of the three would probably find fault with the other two.

Vilas County

Few other parts of the world have as many lakes per square mile as the moraines and pitted outwash plains of Vilas County. These lakes have lots

Fig. 15.9. The commercial establishments in a resort area must be as conspicuous as possible to attract the business on which they depend.

of fish, and they are connected by sluggish streams, so it is easy to move from lake to lake. Splendid stands of white pine first attracted lumbermen, who built logging railroads that made the county an overnight train trip from Chicago for venturesome vacationers and for those who suffered from hay fever and asthma.

By 1900 the logging boom had ended, and groups of businessmen bought large tracts of cutover land that they managed as hunting and fishing preserves. Initially the members of rod and gun clubs roughed it in tents, but soon they built fancy lodges. Less affluent fisherpeople stayed with local families, who developed resorts and rented rustic cabins, many of which lacked running water and were lighted by kerosene lamps or even candles. Vilas County has always boasted a rough-and-ready atmosphere, and it was a favorite gangster hideout during the Prohibition era, when drinking, gambling, and prostitution flourished on an "as if legal" basis even when they were prohibited.

After World War II, regular visitors began building their own summer cottages, and resorts modernized their cabins into housekeeping units; at least one Iowa farm family even brought its own hens to lay its breakfast eggs. Facilities for overnight stays are limited, but the county has an abundance of campgrounds and trailer parks. It also has more than its share of gunsmiths, taxidermists, guide services, and businesses that sell and repair powerboats, snowmobiles, and all kinds of small engines (fig. 15.10). Vilas County has hundreds of miles of snowmobile trails, and on the third weekend of January forty thousand people converge for three days of world-championship snowmobile races.

Door County

The Door peninsula extends seventy miles northeast from the city of Green Bay. The western side has sheer wave-cut cliffs interrupted by shallow pocket harbors that are ideal for pleasure boats. The eastern side has wide expanses of sandy beach. The peninsula was settled largely by Scandinavians who knew how to work with wood. They cut cordwood for sale to lake steamers, developed a brisk trade in cedar shingles for the Chicago market, and left a legacy of many fine old log houses and barns.

Regular steamship service from Chicago lured the first summer visitors. Initially they boarded in private homes, but there was enough business to support summer hotels before World War I. Visitors came by lake steamer because the county never had good rail service. The boat trip from Chicago took three days, with stops at Milwaukee and lesser ports along the way.

Only people with long vacations could afford the three-day trip. Door County attracted college professors from Illinois and Wisconsin and professional people from Chicago and Milwaukee, who built cottages in which they could estivate. They established a sanctuary of peace and quiet where they could commune with nature, cultivate culture, and think deep and beautiful thoughts. The county boasts the oldest summer-stock theater company in the United States, a flourishing summer music festival, and a variety of other cultural delights.

After World War II, the automobile made Door County much more accessible, but the county has tried to retain its character. It has never been comfortable with powerboats or snowmobiles, and vastly prefers sailboats and cross-country skiing (fig. 15.11). The picturesque scenery and quiet lifestyle has attracted artists, who have established permanent studios and galleries in quaint old log buildings, and the county has a plethora of gift shops, antique stores, and cute boutiques to serve casual visitors and busloads of overnight trippers who come to enjoy the scenery. There are camp-

Fig. 15.10. Hunting and fishing resort areas support taxidermists.

grounds and resorts for longer visits, but most of the county's hotels and motels charge daily rates for stays of only a night or two.

Adams County

Adams County has not one but three quite distinct resort areas. The central sand plain is notorious as the poorest area in Wisconsin, and as late as 1960 much of the land in the county could be bought for as little as fifty dollars an acre, or even for back taxes. The county became a haven for people from Milwaukee and Chicago who could not afford more-expensive resort areas. They bought small lots and built their own simple cottages, often without running water or electricity. Many have immobilized mobile homes on their properties, and they have gradually added sundecks, new rooms, garages, toolsheds, and other improvements (fig. 15.12). Even though the structures are modest, they are maintained meticulously, and the owners have done whatever could be done with a great investment of time, effort, and tender loving care but only a modest expenditure of cash.

In 1968 an experienced developer bought a large acreage in Rome

Fig. 15.11.   Sailboats in Door County, Wisconsin.

Township in northern Adams County, impounded three new lakes, and sold two thousand half-acre lakeshore lots on each one. Purchasers of lots on the newest lake become members of a total recreational complex (fig. 15.13), with a handsome three-story clubhouse, two large swimming pools, golf courses, tennis courts, a ski chalet with short downhill runs, and a beach house and boat dock on the lake, which has been stocked with fish. Some lot owners have built summer cottages; others have built permanent homes from which they commute to nearby paper-mill towns. Rome Township had the second-highest rate of population growth in the entire state of Wisconsin in the 1970s.

Just beyond the southwestern corner of Adams County is one of the world's truly awesome family-oriented commercial entertainment area, the Wisconsin Dells. The dells, where the Wisconsin River has carved a narrow gorge with sheer cliffs and grotesque rock pillars, have been a tourist mecca since 1848, although they have been overshadowed by commercial development in recent years. Twenty-five major tourist attractions compete with rides, stage shows, entertainment complexes, motels, and restaurants in using outlandish architecture and garish signs and billboards. The Wiscon-

Fig. 15.12. A camper trailer on a three-acre lot in Adams County, Wisconsin.

sin Dells is the kind of place that aesthetes deplore but people adore; more than two and a half million people visit the area each year.

SKIS AND SNOWMOBILES

From Barrel Staves to Bognors

Skiing, the most popular outdoor winter sport in the United States, did not really catch on in this country until 1933, when an ingenious Yankee mechanic in Woodstock, Vermont, hitched a stout rope to an old Model T Ford engine and invented the first rope tow to haul enthusiasts back uphill after they had skidded their way to the bottom. The railroads started running special weekend ski trains from eastern cities to nearby ski areas, and a few ski resorts were developed in the West, but these activities were halted by the war, and the big ski boom did not come until after World War II.

Ski areas are widely scattered through the North and the West, but the greatest numbers are near the major cities of the Northeast (fig. 15.14). Weekend ski areas, which have limited facilities and draw their customers

Fig. 15.13. The clubhouse on a new recreational development in Adams County, Wisconsin.

mainly from nearby cities, differ markedly from the major destination ski resorts, which have better conditions and facilities for skiing but must draw their clientele from much greater distances.

A successful weekend ski area must have a chalet or central lodge with a bar and restaurant, a shop where equipment can be bought or rented, a snowmaking machine, one or more lifts, a suitable parking area, and good access roads; it helps to have lights so the slopes can be used at night. The machines for making artificial snow, which have been developed since World War II, have been an enormous boon to weekend ski areas, and have permitted their extension southward as far as the Great Smoky Mountains of western North Carolina and eastern Tennessee. The nozzles of the machine spray a mist of fine water droplets across the slope, and these turn to snow before they settle to the ground if the air is cold enough. The machines will work at 32°F, but the ideal is a temperature of 28°F or lower, with low humidity.

Major destination ski resorts are in areas where longer slopes and heavier snows provide a variety of skiing conditions. They must also have luxurious accommodations and lively entertainment to attract and hold their visitors. They need easy access by car and plane, large, modern lodges, fast and efficient lifts, a selection of bars and restaurants, plenty of shops and boutiques, and good medical facilities. It helps to have facilities for other winter sports, such as skating rinks and bobsled runs, and most ski resorts have attempted to extend their seasons by developing golf courses, swimming pools, tennis courts, riding trails, artificial lakes, and convention facilities. They have generated local real estate and building booms for condominium apartments and vacation homes, and they face the same problems as other contemporary urban areas, including traffic jams, air pollution, and the need for planning, building codes, and sewage plants.

The Snarling Snowmobile

The snowmobile erupted across the winter countryside of the North American northlands in the 1960s, and it revolutionized wintertime outdoor recreation. The number of the low-slung, highly maneuverable little vehicles in use jumped from fewer than two thousand in 1959 to half a million in 1969, and it has continued to increase. The machines are aimed by handlebars connected to a pair of steel skis in front, and propelled at speeds up to sixty miles an hour by a rear-mounted cleated rubber belt driven by a two-cycle air-cooled engine that sounds like a demented chainsaw.

The snowmobile has been a godsend for those who must live with winter in the northlands. It enables them to get to places that previously they could reach only on snowshoes or skis, and then with difficulty. Utility workers can service lines even when the roads have been blocked by drifting snow, and foresters can continue to cruise timber through the winter. Ranchers can haul feed to their starving cattle, ride their fences, and make repairs. No longer need they fear being snowbound, because in emergencies their snowmobiles can carry them out or bring in the doctor they need. The trapper can patrol a longer trapline, and even the Inuit have traded in their dog teams for snowmobiles. The crowning indignity came in 1969, when the Royal Canadian Mounted Police replaced the last of their fabled sled dogs with the new machines.

Snowmobiles have opened up new vistas for outdoor recreation in winter. They buzz over the snow-covered slopes of golf courses, tow sleds and skiers, and haul fishing shanties out onto frozen lakes. Many of these shanties have all the comforts of home—heaters, carpets, easy chairs, and refrigerators full of liquid refreshments. Some years ago four enterprising women in St. Paul, Minnesota, made the front pages when the police dis-

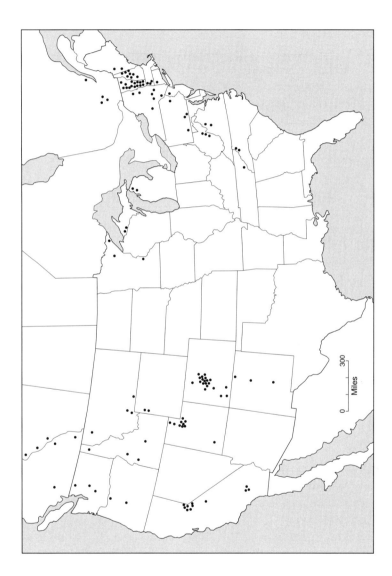

Fig. 15.14. Major ski resorts, 1996–97.

covered that they had set themselves up for business in a shanty on one of the larger lakes nearby and were plying quite a profitable trade.

Snowmobile enthusiasts formed clubs and organized expeditions, and some of the hardier types even experimented with winter camping. One group set off for the North Pole on their trusty snowmobiles and actually made it. Resort and summer-cottage owners began to keep their places open well into the winter, and managers of state and national forests developed trails exclusively for snowmobilers. On the other hand, the little vehicles have a perverse habit of tipping over, getting stuck, breaking down, crashing through thin ice, leaving litter, breaking through fences, mowing down young trees, frightening animals, and rousing the countryside with their infernal racket. They have not been universally well received.

CASINOS

Although they can hardly be described as outdoor activities, gambling casinos on Indian reservations are one of the newest types of rural recreational facility in the United States. Casinos have played a major role in the economic growth of some metropolitan areas, such as Las Vegas and Atlantic City. They boomed near Tunica south of Memphis and on the Gulf Coast of Mississippi after that state relaxed its gambling laws in 1990, and a number of other cities seem to have turned from new industry to casinos in their search for a panacea for their economic ills.

The development of new casinos in rural areas dates from 1988, when Congress passed the Indian Gaming Regulatory Act, which allows tribes to operate any gambling activities on their reservations that are permitted in their states, and by 1994 there were at least 143 reservation casinos in twenty-four states.

In some states these reservation casinos are little more than bingo parlors with slot machines, but some are extremely impressive. The Foxwoods Casino, twenty miles northeast of New London, Connecticut, was opened by the Mashantucket Pequot Indians in 1992; it is the largest casino in the United States and the most profitable in the world, with annual revenues of more than one billion dollars. Foxwoods is larger than five football fields, with 4,428 slot machines, 308 table games, and a three-thousand-seat bingo hall.

Minnesota and Wisconsin have the greatest concentration of reservation casinos, with a total of thirty-two, and Indian gaming has become the seventh-largest industry in Minnesota. Most reservations are in inaccessible rural backwaters with poor roads, and their casinos are fairly modest affairs, but the Mystic Lake Casino near Minneapolis rivals anything Las Vegas has to offer.

Mystic Lake is surrounded by an enormous parking lot for thousands of vehicles, with special areas for tour buses, stretch limousines, recreational vehicles, and snowmobiles. Down the center of the vast interior are rows of semicircular blackjack tables, and around the edges are batteries of slot machines, whose insistent drone and clatter create the distinctive casino sound. In 1994 Mystic Lake had 128 blackjack tables and 2,350 slot machines. The soft glitter of the casino contrasts sharply with the harsh glare of the adjacent bingo parlor, which is the size of a football field.

The clientele is strongly blue-collar. Everyone seems deadly serious, and no one ever smiles, yet for many people the casino clearly is an attractive alternative form of entertainment. The slot machines have payoff rates as high as 97 percent, and an afternoon at the casino need cost no more than a concert, a baseball game, or a movie. The total sums involved are staggering. In 1992 the Federal Reserve Bank of Minneapolis had a sudden severe shortage of quarters, presumably because of the demand from the newly opened casinos. It placed an emergency order with the United States Mint for eight million dollars' worth of quarters, whose delivery required a fleet of ten semitrailers.

Casino income has given many Indians an improved sense of self-worth, and tribes have invested their profits in better housing, improved roads, new schools and health-care clinics, and day-care and community centers. Few tribes make direct payments to their individual members, but in 1993 the band that operates Mystic Lake, which is far and away the most profitable in Minnesota, paid half a million dollars to each full member.

The new prosperity of the tribes has aroused the envy and animosity of some of their neighbors, and the owners of bars, restaurants, and bowling alleys have pressured the legislature to be allowed to have their own slot machines. The tribes realize that competition is inevitable and they are investing in nongambling enterprises. Concern has also been expressed about the effects of sudden wealth on poor people, about the possible infiltration of the casinos by organized crime, and about the danger that some customers will become compulsive gamblers. The casinos could not be making all that money if someone were not losing it, and for some patrons the casinos are deadly traps rather than places for pleasant entertainment.

Gambling casinos on Indian reservations are merely the most recent manifestation of the desire of American city folk to seek recreational opportunities in rural areas. Rural areas that boast amenities, however defined—topography or water or deep snow or balmy winters or appealing plant life—have been the only rural areas in the United States that have been gaining population, and they have spawned a host of commercial entertainments, ranging from theme parks to casinos, to attract ever-greater numbers of visitors.

Epilogue *The Changing Countryside*

Rural landscapes are the creatures of human activities. Over the centuries people have cleared the forests, busted the sod of the grasslands, and drained the swamps and marshes. They have divided the land into political units and private properties. They have erected fences to mark the boundaries of their fields and pastures, and they have built shelters for their crops and animals.

They have founded central places to collect, process, and market their products and to provide the services they need. They have dug minerals from the earth and built mills and factories in which to process them. They have developed networks of highways and waterways to facilitate the flow of people, goods, and ideas. They have come to appreciate the value of the countryside for recreation, and they have come full circle by replanting some of the forests their ancestors destroyed.

Two centuries ago the United States was an agricultural nation. In 1790 only two of every forty Americans lived in a town or city, but in 1990 only one of every forty lived on a farm. Most contemporary Americans are removed at least a generation or two from any firsthand knowledge of farming, which has changed dramatically during their lifetimes. Many of the old familiar structures of the agricultural landscape have been greatly modified or demolished because technology has made them obsolete, or farm consolidation has made them redundant, or farmland abandonment has made them derelict.

One of the most important catalysts for change in rural areas during the twentieth century has been the internal-combustion engine. In its various manifestations it has transformed or outmoded not only traditional farming systems but also business districts, settlement patterns, and political processes. A major challenge facing contemporary society has been to try to figure out how to adapt obsolescent patterns and activities to changing human needs, wishes, and behavior. What can be rescued, what can be recycled and rehabilitated, and what is so obsolete that it must be jettisoned and replaced?

FARMING SYSTEMS

The tractor (along with its permutations, such as the self-propelled combine) has replaced the horse. It has greatly reduced the need for farm labor, thus freeing or forcing hordes of farm youths to seek city jobs, and it has greatly increased the minimal size necessary for a viable farm operation.

Farming today, like manufacturing in the seventeenth century and the contemporary grocery business, is in the process of reorganizing into larger and more-efficient units. Most modern farms are already too small, and the minimal size for a successful modern farm operation is steadily increasing. Farmers whose farms are undersized may become part-time farmers, or they may become part-owner farmers.

Part-time farmers find off-farm jobs, work eight hours a day in town, use their morning, evening, and weekend hours to work on the farm, and time their vacations from their off-farm jobs for the periods when the workload on the farm is heaviest. After a while this ceaseless routine of work begins to get old, and eventually part-time farmers decide to lease their land to part-owner farmers.

Part-owner farmers continue to farm the land they own, and they enlarge their operations by renting land from owners who no longer wish to farm it. They have to rent whatever they can find in a tightly competitive market, so most part-owner farms consist of widely scattered tracts of land, despite the obvious inefficiencies of such a layout. The old farmstead on a rented farm often is redundant, and many former farmsteads have been razed, but some old farmhouses have been recycled into rural residences for city people who want to live in the country.

Part-ownership and farm consolidation have increased the size of farms, and the number of farms has dwindled as their size has increased. The number of census "farms" in the United States dropped from nearly seven million in 1934, the peak year, to only two million in 1992, but the five million "lost" farms, as well as more than half of those that still remained on the list, were really nonfarm farms that were undersized for modern farm operations.

Larger farms demand more-skillful professional management, and brainpower has become as important as muscle power on modern farms. Farming has changed from a way of life into a complex business that demands a wide range of management skills, including the skill of money management. Modern farmers must be able to handle large amounts of money just as competently and efficiently as they handle powerful tractors and other large farm machines, and the old family farm has had to become an efficient family farm business.

Major technological innovations have enabled farmers to increase their productivity dramatically. In 1990 each hour of farm labor produced more than three times as much food and fiber as in 1950. Chemists have given farmers a whole new arsenal of agrichemicals to help them reduce losses to weeds, insects, and diseases, and to improve the productivity of their soil. Engineers have designed vastly improved farm equipment, such as better machines to pick tomatoes, while plant breeders have developed new tomato varieties that are easier for machines to pick.

Plant and animal breeders have scored impressive breakthroughs. The story of hybrid corn is best known, but entomologists have managed to screw up the sex lives of some insect pests so completely that the poor little things are no longer able to breed, and the marvels of biotechnology still remain just tantalizingly over the horizon.

Today's successful farmers must concentrate their energies and their resources by specializing in the intensive large-scale production of the few commodities that their computers tell them they can produce most competitively. The old general farm that produced a little bit of everything is as dead as the dodo. Some farmers specialize in growing crops, others specialize in feeding livestock, and both crop and livestock farmers buy their milk and groceries at the local supermarket, just as you and I do.

Specialized crop farmers have transformed the rural landscape by enlarging their fields, pulling out their fences, and replacing their traditional farm buildings with newer structures that are better suited to the larger operations. These farmers have largely jettisoned traditional rotations, and are concentrating on producing the crops that are most remunerative.

Specialized livestock production is shifting to a larger scale and becoming more industrialized. Producers in the West have developed huge dry lots for beef and dairy cattle. Cattle farmers in the East have followed the lead of the dry-lot operators, and now they buy much of their feed instead of trying to grow their own. Dry lots are common even in the dairy country of Wisconsin, where farmers harvest their forage crops with machines precisely when they are ripe, and store them as silage or hay until the cattle need them. No pasture is wasted because the animals have trampled or soiled it, and many dairy farmers are removing the fences they no longer need.

Hog and poultry production also has become highly specialized, and long, low, one-story, purpose-built houses for hogs and poultry have become distinctive features on the landscape. Vertical integration originated in the broiler business. The integrator provides chicks and feed, the farmer feeds the birds to market weight, and the integrator processes and sells them. The turkey business has followed the lead of the broiler business, and today most of our eggs are laid in huge "factories" in which millions of hens are doing their daily duty.

Hog farming is also shifting toward large, specialized, industrialized operations that produce hundreds of thousands of nearly identical hogs a year in climate-controlled buildings. Small-scale hog farmers are threatened by the huge new operations, and traditional hog-farming states have passed laws restricting them, so pork-processing companies have shifted their operations to states such as North Carolina and Oklahoma, where the regulatory environment is less restrictive.

Greater specialization and the greatly increased volume of production require greater skill in seeking and serving wider markets. The motor truck (and later the jet airplane) have replaced the horse and the farm wagon, and they have globalized the market for farm products. Farmers in Iowa now look nervously at Brazil before they decide to plant soybeans, and farmers in Florida and California fret that the free-trade agreement with Mexico will put them out of the tomato business.

A jumbo jet loaded with flowers takes off from Tel Aviv for Amsterdam every other winter day, and on Valentine's Day a florist in Minneapolis sells fresh flowers flown in from Mexico, Costa Rica, Colombia, Ecuador, Holland, and Thailand. The leading meat-packing company in Japan, after exploring eleven different countries, decided to develop a hog operation in the Texas panhandle that will produce half a million hogs a year and ship the pork to Japan.

Farms of family size, with sales of one hundred thousand to five hundred thousand dollars, still dominate most of the eastern United States, but specialized farmers in California, Arizona, and Florida have had to develop large-scale operations to produce high-value products that can bear the cost of transporting them long distances, and they have had to develop sophisticated systems for marketing them (fig. E.1).

Greater specialization has permitted greater selectivity in the use of

land. Agricultural production has become concentrated on the best farmland, and farmers on poorer land are no longer competitive. Some continue to farm in a desultory, part-time fashion, but much marginal agricultural land has been abandoned and allowed to grow up in weeds, briers,
and brush.

The very notion of farmland abandonment is anathema to most modern Americans, who still think of land in terms of clearance, not abandonment. The hardy pioneers who felled the forests and brought the land into
cultivation are among our national folk heroes, and we feel that somehow
we have let them down when we discover that an area larger than the entire
state of Iowa has gone out of farming during the twentieth century. It should
have.

Most of the eastern United States will revert to woodland within fifty
years or so unless people do something to prevent it, and people have not
seen fit to prevent it along the spine of Appalachia, in that broad belt of
rugged hills and mountains that runs northeastward from Alabama into
northern Maine and beyond. The land is too steep and stony for modern
farm machinery, and people have allowed nature to foreclose its mortgage
on it. Even nature, however, can do only so much to heal land that has
been worn and battered by two centuries of human use and abuse, and the
forests it is putting back on the land are much inferior to those that originally cloaked it.

SMALL TOWNS

The automobile freed people to shop where they wished when it replaced the horse and buggy. They could bypass small towns for larger places
that offered a greater variety of goods and services at lower prices. Today
most small towns look dead. Main Street indeed is dead, because the former retail and service functions of small towns have shifted up the urban
hierarchy to larger places. Small towns also lost their function as collecting
and marketing centers for the farming areas around them when the motor
truck replaced the horse and the farm wagon.

Despite these losses, the population of many small towns continues to
increase in fitful fashion. Their houses are inexpensive homes for people
who are footloose or retired, and local entrepreneurs may have developed
business establishments that are disproportionate to the size of the place.
Many small towns have found new niches as minor cogs in the national

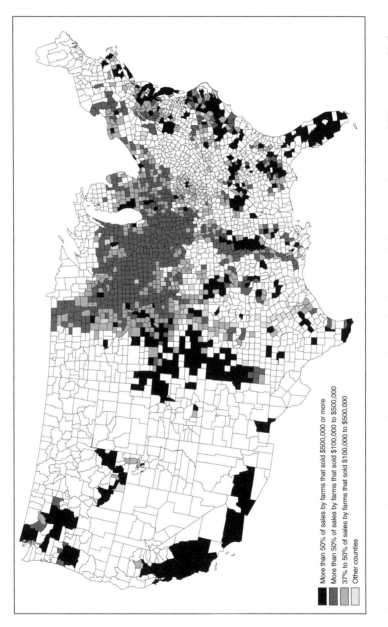

Fig. E.1. The source of sales of farm products in 1992, in counties that exceeded the national average of $72 an acre, with farms classified by their volume of sales. The West, including the central and southern Great Plains, is dominated by megafarms with sales of half a million dollars or more. The Midwest is dominated by family farms with sales of one hundred thousand to five hundred thousand dollars. The South is a crazy quilt, except for the megafarms of Florida, and the Northeast has relatively few farms of any size.

More than 50% of sales by farms that sold $500,000 or more

More than 50% of sales by farms that sold $100,000 to $500,000

37% to 50% of sales by farms that sold $100,000 to $500,000

Other counties

system of manufacturing centers. Their "factories" often do not look like factories, because many use floor space in recycled older buildings, such as redundant schoolhouses or stores.

Small towns are learning that they must cooperate in order to survive. A group of small towns can support establishments larger than any one of them alone could support, and each place can have its fair share of the establishments that serve the entire group. A group of small towns can cooperate to form a dispersed city, and a group of cities can cooperate to form a dispersed metropolis.

The automobile has also changed the business districts of larger places, which have been turned inside out. The new downtown is now at the edge, and the old central business district is little more than a convenience shopping area for those who work there during the day. After dark it is virtually deserted. It has been superseded by a new business district that does not yet even have a generic name. This new business district has variously been called the suburban freeway corridor, the bypass strip, and the edge city.

The new business district is an endless strip of office buildings, motels, filling stations, fast-food joints, and stores set in a cornucopia of parking lots and anchored by shopping malls. Intellectuals have been told that they should dislike it, and they dutifully obey orders, but for many ordinary Americans it is the highest and best contemporary achievement of the American way of life. It is a place where they can feel at home, no matter where they happen to be, because it is so familiar, so standardized and universal, so placeless. It seems like home to strangers, and it keeps traveling Americans from feeling like strangers no matter where they may roam.

The new business district welcomes automobiles, and in so doing it reflects the changes that technology has wrought in American society. The automobile has joined the house as a prime symbol of basic values, and nothing tells quite so much about Americans as the houses they choose to live in and the cars they choose to drive.

RURAL RESIDENCE AND RECREATION

The internal-combustion engine has helped transform settlement patterns in rural areas. The tractor greatly reduced the need for farm labor, and the automobile enabled the surplus agricultural labor force to seek and take city jobs. It also divorced the homes of workers from their workplaces, and enabled hordes of city people to invade the countryside in search of residence and in search of recreation.

The automobile has greatly extended the distance a commuter can travel, and fifty miles no longer seems an onerous journey to work. Furthermore, many of the best new jobs are in the new downtown at the city's edge, and they are as accessible from rural areas as from the older, built-up areas of the city.

It is tough to tell where the city ends and the countryside begins. The traditional distinction between urban and rural, between city slicker and country bumpkin, has completely lost what little meaning it might once have had, and the idea of a neat and tidy rural/urban dichotomy is an outmoded, trivial holdover from the horse-and-buggy era.

The traditional concept of rurality was a bundle of many different traits. At one time these traits were tightly interrelated, but no more. Once you could use any single trait to identify the entire bundle, but the fumes of the internal-combustion engine have dissolved the ties that once bound this bundle of traits together, and the concept of rurality has lost whatever analytical power it once had.

Although it has lost its analytical power, the concept of rurality still remains useful, because vernacular speech needs fuzzy words such as *rural, landscape, environment, sustainable,* and *place.* We all think we understand these words, but they lose their usefulness if we try to define them precisely enough to be useful for analytical purposes. For example, we can all identify the ends of the rural/urban continuum, but trying to draw a dividing line between rural and urban is simply a waste of time.

The residential tentacles of the modern metropolis reach for miles into the countryside, especially in areas where farmland is not too expensive and planning is permissive. The most advanced form of this new dispersed city has evolved spontaneously on the Piedmont of the Carolinas and Georgia and in the Great Valley of eastern Tennessee and southwestern Virginia, where every paved road seems to be lined with nonfarm houses.

This area has no name, so John Morgan and I have called it Spersopolis, the dispersed city. The highwayside strips of Spersopolis might well represent the residential component of the new populist ideal city that Americans would build if they could. The new downtown at the city's edge is the principal business district of this new city.

The houses of Spersopolis are as close to the highway as possible, for easy commuting. They screen off the countryside behind them, and those who stick to the main highways, especially the main highways between major cities, see little more than the screen. Back of the screen, if they have the courage to explore, they will find lots of open country that is still used for farming.

Preposterous though the idea might seem to denizens of Los Angeles or Long Island, the cities in this country are not about to take over our farmland. It is astonishing that the United States has no accurate information on the extent of its built-up area, but the best estimate is that cities, towns, villages, hamlets, airports, highways, and all other built-up areas cover no more than about a hundred million acres, or only 5 percent of our total land area.

We do have accurate statistics on our cropland acreage, and we have been losing about a million acres a year, but some of it is no longer cultivated because it is pretty sorry land. Even a million acres is only one-half of one percent of the nation's base of 280 million acres of cropland harvested. At this rate of loss we can start worrying about a cropland shortage around the year 2150, which is still fairly far down the pike for most of us.

Some city people come to rural areas in search of residences, but larger numbers flock to the countryside in search of land for recreation, for second homes, and for retirement. People will travel considerable distances to visit commercial recreational areas such as Disneyland and Disney World, the Wisconsin Dells, Gatlinburg, Tennessee, or Branson, Missouri, but they rarely stay longer than a week or so. Bodies of water are the major rural recreational magnets in the summer months, but hill and mountain areas with cool weather and scenic topography also attract many summer visitors.

Recreation is a highly seasonal activity, and the local people have less than four months in which to milk the tourists of enough to live on for the rest of the year. Most tourist areas have done their best to expand into the spring and fall "shoulder" seasons. In winter the most popular outdoor activity for adults in white-collar resort areas is skiing. The snowmobile, wintertime cousin to the speedboat, is more popular in blue-collar areas, which attract people who want speed and blood rather than peace and quiet. In white-collar areas people watch birds; in blue-collar areas they shoot them.

The ever-greater numbers of city people invading the countryside can be a mixed blessing. An excessive number of visitors can debase the quality of the very amenities they seek to enjoy, but increased exposure of city people to the countryside can also intensify their concern for protecting and preserving it.

City people who live in rural areas can help maintain the beauty of the countryside by spending their money to keep it attractive, but most working farmers can no longer afford that luxury. In fact, some people have seriously proposed that farmers should be paid salaries as museum curators, to serve as the custodians and caretakers of an open countryside to which city folk can enjoy pleasant outings.

Most city folk who own land in the country can do their bit. By far the most impressive examples are showplace "gentleman farms." Nearly every major city has a few, but the greatest concentration is in the Estate Belt just west of Megalopolis. Gentlemen farmers like to gaze out on well-kept fields and pastures and fine livestock. Their ability to pay for what they like has created some of the most beautiful rural landscapes in the United States.

POLITICAL PROCESSES

Automobility has enabled contemporary society to become ever more metropolitan, ever more divorced from the land, but it also enables hordes of metropolitan people to invade the countryside in search of residence and recreation. The invaders from the city often insist on changes that country people find unpalatable, and conflict seems almost unavoidable. The most-intense conflict is at the rural/urban fringe, where the built-up area of the city is advancing inexorably into rural areas.

The rural/urban fringe is the new frontier of modern metropolitan society, ruled by the power of the purse. Money determines what happens on the rural/urban fringe, where developers are able to convert huge chunks of rural land to housing areas, shopping malls, office parks, and other urban uses, and those who object seem powerless to halt the economic juggernaut.

One of the principal functions of government is to protect society against the economic excesses of the market. Every polity has evolved an institutional framework for mediating conflict and allocating precious land, but existing institutions on the rural/urban fringe often are overwhelmed by the speed of change and the economic power of developers, and they are incapable of managing and controlling development.

Those who dislike development on the modern metropolitan frontier have to become involved in the political process. They must learn to work the levers of political power, because they lack the economic muscle necessary to forestall or control development. They must arouse and mobilize the lethargic body politic, which is no easy task. At times they have felt compelled to overstate their case in order to attract support for their cause, and in so doing they have seriously damaged their credibility.

Some critics of development also have camouflaged their objectives by using catchy codewords. Who, for example, could possibly be antienvironmentalist? Who could oppose "sustainability," which became a popular catchword in the early 1990s? No one can object to sustainability because no one really knows what it means. Some critics say that this slogan seems to mask a rejection of much that is modern, and a desire to turn back the clock to an earlier age that is perceived to have been purer, wiser, and better.

Sloganeering is not an acceptable substitute for clear, logical, analytical thought. Enlightened decision making must be based on a good understanding of contemporary society and of the processes that have shaped and are shaping it, not on sentiment, no matter how well intentioned. We cannot cast wise votes about what we should be doing and where we should be going until we understand where we are and how we got here.

More and more people believe that understanding should lead to action. They argue that application is implicit in any analysis, that we should use our understanding to make good predictions, and that we should act on our predictions. Others remember the dictum that those who live by the crystal ball must learn to eat broken glass. They are content to make the effort to understand, and are willing to leave prediction and the politics of implementation to those who are so inclined, but they do agree that scholars have a major obligation to help activists understand the processes they are trying to influence.

Society needs scholars, and society needs advocates, but they have oppositional mind-sets. Scholars think; advocates believe. The advocate makes the best possible case for a preconceived conclusion by emphasizing favorable evidence and ignoring or suppressing all evidence to the contrary. The scholar strives for a conclusion that incorporates every possible shred of evidence, and has no preconceived idea about what that conclusion will be, much less what it ought to be.

A FINAL WORD

The rural landscape is constantly changing. It is always in a state of becoming. We must understand what it was in order to understand and appreciate what it is and what it will become. It is shaped by a host of individual decisions, some great and transformational, most small and incremental in and of themselves. In the aggregate, however, many small decisions can add up to enormous change.

The rural landscape no longer belongs solely to farmers, if indeed it ever did. Larger and larger numbers of city people are claiming a right to use the countryside, and conflict over its right and proper use will inevitably increase.

The major features of the earth's surface seem constant in human (albeit not in geological) time, but people have demonstrated remarkable ingenuity in shaping and molding the lesser features to serve their needs.

The uncultivated plant life that seems so natural has actually been modified by multitudinous human actions, and the cultivated plant life clearly reflects the needs and wishes of those who cultivate it.

The structures that people have added to the rural landscape reflect the needs and the values of those who are its creators, its inhabitants, and its custodians, and we must learn to see it through their eyes to understand and appreciate it fully.

Further Reading

CONCEPTS

Hart, John Fraser. "The Highest Form of the Geographer's Art." *Annals of the Association of American Geographers* 72 (1982): 1–29, especially 7–8.

Hartshorne, Richard. "'Landschaft' and 'Landscape.'" In *The Nature of Geography*, 149–74. Washington, D.C.: Association of American Geographers, 1939.

Meinig, D. W., ed. *The Interpretation of Ordinary Landscapes: Geographical Essays.* New York: Oxford University Press, 1979.

Sauer, Carl O. *The Morphology of Landscape.* University of California Publications in Geography 2, no. 2 (1924).

Thompson, George F., ed. *Landscape in America.* Austin: University of Texas Press, 1995.

THE SURFACE OF THE LAND

Atwood, Wallace W. *The Physiographic Provinces of North America.* New York: Ginn, 1940.

Fenneman, Nevin M. *Physiography of Eastern United States.* New York: McGraw-Hill, 1938.

———. *Physiography of Western United States.* New York: McGraw- Hill, 1931.

Finch, Vernor C., and Glenn T. Trewartha. *Physical Elements of Geography.* 3d ed. New York: McGraw-Hill, 1949.

Hamblin, W. Kenneth. *Earth's Dynamic Systems: A Textbook in Physical Geology.* New York: Macmillan, 1989.

James, Preston E. Appendix C, "The Lithosphere." In *A Geography of Man*, 3d ed., 489–516. Waltham, Mass.: Blaisdell, 1966.

Lobeck, A. K. *Geomorphology.* New York: McGraw-Hill, 1939.

Shelton, John S. *Geology Illustrated.* San Francisco: Freeman, 1966.

Thornbury, William D. *Regional Geomorphology of the United States.* New York: Wiley, 1965.

LANDSCAPES OF MINING

Jackson, R. T. "Mining Settlements in Western Europe: The Landscape and the Community." In *Urbanization and Its Problems: Essays Presented to E. W. Gilbert*, edited by R. T. Beckinsale and J. M. Houston, 143–70. Oxford: Basil Blackwell, 1968.

PLANT LIFE

Braun, E. Lucy. *Deciduous Forests of Eastern North America*. Philadelphia: Blakiston, 1950.

Krebs, Charles J. *Ecology: The Experimental Analysis of Distribution and Abundance*. New York: Harper and Row, 1979.

Küchler, A. W. *Manual to Accompany the Map: Potential Natural Vegetation of the Conterminous United States*. Special Publication no. 36. New York: American Geographical Society, 1964.

Millers, Imants, David S. Shriner, and David Rizzo. *History of Hardwood Decline in the Eastern United States*. General Technical Report NE-126. Broomall, Pa.: Northeastern Forest Experiment Station, 1989.

Silvics of North America. Agriculture Handbook 654. Washington, D.C.: U.S. Department of Agriculture, 1990.

Watts, May Theilgaard. *Reading the Landscape: An Adventure in Ecology*. New York: Macmillan, 1957.

Whittaker, R. H. *Communities and Ecosystems*. New York: Macmillan, 1970.

PLANTS AND PEOPLE

Dingman, Charles E. "Land Alienation in Houston County, Minnesota: Preferences in Land Selection." *Geographical Bulletin* 4 (1972): 45–49.

Hart, John Fraser. "Land Use Change in a Piedmont County." *Annals of the Association of American Geographers* 70 (1980): 492–525.

Kiefer, Wayne E. *Rush County, Indiana: A Study in Rural Settlement Geography*. Bloomington: Indiana University Department of Geography, 1969.

May, Dennis M. *Development and Status of Arkansas' Primary Forest Products Industry*. Resource Bulletin SO-152. New Orleans: Southern Forest Experiment Station, 1990.

Reaman, G. Elmore. *The Trail of the Black Walnut*. Toronto: McClelland and Stewart, 1957.

Stokes, George A. "Lumbering and Western Louisiana Cultural Landscapes." *Annals of the Association of American Geographers* 47 (1957): 250–66.

Swanholm, Marx. *Lumbering in the Last of the White Pine States*. Minnesota Historic Sites Pamphlet Series, no. 17. St. Paul: Minnesota Historic Society, 1978.

Williams, Michael. *Americans and Their Forests: A Historical Geography*. Cambridge: Cambridge University Press, 1989.

CROPPING SYSTEMS

Hart, John Fraser. "Land Rotation in Appalachia." *Geographical Review* 67 (1977): 148–66.

———. "Loss and Abandonment of Cleared Farm Land in the Eastern United States." *Annals of the Association of American Geographers* 58 (1968): 417–40.

Leighty, Clyde E. "Crop Rotation." In *Soils and Men*, 406–30. Yearbook of Agriculture, 1938. Washington, D.C.: U.S. Department of Agriculture, 1938.

Mather, Eugene, and John Fraser Hart. "The Geography of Manure." *Land Economics* 32 (1956): 25–38.

LAND DIVISION IN BRITAIN

Cole, Sonia. *The Neolithic Revolution*. London: British Museum, 1961.

Evans, E. Estyn. *Irish Folkways*. London: Routledge and Kegan Paul, 1957.

———. *Irish Heritage: The Landscape, the People, and Their Work*. Dundalk, Ireland: W. Tempest, 1942.

———. *Mourne Country: Landscape and Life in South Down*. Dundalk, Ireland: Dundalgan Press, 1967.

Hoskins, W. G. *The Making of the English Landscape*. London: Hodder and Stoughton, 1955.

Orwin, C. S., and C. S. Orwin. *The Open Fields*. 3d ed. London: Oxford University Press, 1967.

Rackham, Oliver. *The History of the Countryside*. London: J. M. Dent and Sons, 1986.

Taylor, Christopher. *Fields in the English Landscape*. London: J. M. Dent and Sons, 1975.

———. *Village and Farmstead: A History of Rural Settlement in England*. London: George Philip, 1983.

LAND DIVISION IN AMERICA

Burghardt, Andrew F. "The Settling of Southern Ontario: An Appreciation of the Work of Carl Schott." *Canadian Geographer* 25 (1981): 75–93.

Gerlach, Russel L. *Immigrants in the Ozarks: A Study in Ethnic Geography.* Columbia: University of Missouri Press, 1976.

Harris, Richard Colebrook. *The Seigneurial Regime in Early Canada: A Geographical Study.* Madison: University of Wisconsin Press, 1966.

Hart, John Fraser. "Field Patterns in Indiana." *Geographical Review* 58 (1968): 450–71.

Hibbard, Benjamin Horace. *A History of the Public Land Policies.* New York: Macmillan, 1924.

Kiefer, Wayne E. *Rush County, Indiana: A Study in Rural Settlement Geography.* Bloomington: Indiana University Department of Geography, 1969.

Marschner, F. J. *Land Use and Its Patterns in the United States.* Agriculture Handbook 153. Washington, D.C.: U.S. Department of Agriculture, 1959.

McIntosh, Charles Barron. *The Nebraska Sand Hills: The Human Landscape.* Lincoln: University of Nebraska Press, 1996.

———. "Patterns from Land Alienation Maps." *Annals of the Association of American Geographers* 66 (1976): 570–82.

Paullin, Charles O. *Atlas of the Historical Geography of the United States.* Edited by John K. Wright and published jointly by the Carnegie Institution of Washington, D.C. and the American Geographical Society of New York, 1932.

Price, Edward T. *Dividing the Land: Early American Beginnings of Our Private Property Mosaic.* Geography Research Paper 238. Chicago: University of Chicago Press, 1995.

Thrower, Norman J. W. *Original Survey and Land Subdivision: A Comparative Study of the Form and Effect of Contrasting Cadastral Surveys.* Association of American Geographers Monograph no. 4. Skokie, Ill.: Rand McNally, 1966.

Trewartha, Glenn T. "Types of Rural Settlement in Colonial America." *Geographical Review* 36 (1946): 568–96.

Vincennes, Indiana–Illinois. 1:62,500 topographic quadrangle. Washington, D.C.: U.S. Geological Survey, 1913.

FENCES

Dowdeswell, W. H. *Hedgerows and Verges.* London: Allen and Unwin, 1987.

Hart, John Fraser. "Field Patterns in Indiana." *Geographical Review* 58 (1968): 450–71.

Hart, John Fraser, and John T. Morgan. "Spersopolis." *Southeastern Geographer* 35 (1995): 103–17.

Mather, Eugene Cotton, and John Fraser Hart. "Fences and Farms." *Geographical Review* 44 (1954): 201–23.

Mead, W. R. "The Study of Field Boundaries." *Geographische Zeitschrift* 54 (1966): 101–17.

Muilenberg, Grace, and Ada Swineford. *Land of the Post Rock: Its Origins, History, and People*. Lawrence: University Press of Kansas, 1975.

Murray-Wooley, Carolyn, and Karl B. Raitz. *Rock Fences of the Bluegrass*. Lexington: University Press of Kentucky, 1992.

Teather, Elizabeth Kenworthy. "The Hedgerow: An Analysis of a Changing Landscape Feature." *Geography* 55 (1970): 146–55.

Webb, Walter Prescott. *The Great Plains*. Boston: Ginn, 1931.

BARNS

Brunskill, R. W. *Illustrated Handbook of Vernacular Architecture*. London: Faber and Faber, 1971.

Burden, Ernest E. *Living Barns: How to Find and Restore a Barn of Your Own*. Boston: New York Graphic Society, 1977.

Comeaux, Malcolm M. "The Cajun Barn." *Geographical Review* 79 (1989): 47–62.

Endersby, Elric, Alexander Greenwood, and David Larkin. *Barn: The Art of a Working Building*. New York: Houghton Mifflin, 1992.

Ennals, Peter M. "Nineteenth-Century Barns in Southern Ontario." *Canadian Geographer* 16 (1972): 256–70.

Ensminger, Robert F. *The Pennsylvania Barn: Its Origin, Evolution, and Distribution in North America*. Baltimore: Johns Hopkins University Press, 1992.

Fitchen, John. *The New World Dutch Barn: A Study of Its Characteristics, Its Structural System, and Its Probable Erectional Procedures*. Syracuse, N.Y.: Syracuse University Press, 1968.

Glass, Joseph W. *The Pennsylvania Culture Region: A View from the Barn*. Ann Arbor, Mich.: UNI Research Press, 1986.

Glassie, Henry. "The Old Barns of Appalachia." *Mountain Life and Work* 41, no. 2 (1965): 21–30.

———. *Pattern in the Material Folk Culture of the Eastern United States*. Philadelphia: University of Pennsylvania Press, 1968.

Hart, John Fraser. "On the Classification of Barns." *Material Culture* 26, no. 3 (1994): 37–46.

———. "Types of Barns in the Eastern United States." *Focus* 43, no. 1 (1993): 8–17.

Hubka, Thomas C. *Big House, Little House, Back House, Barn: The Connected Farm Buildings of New England*. Hanover, N.H.: University Press of New England, 1984.

Noble, Allen G., and Gayle A. Seymour. "Distribution of Barn Types in Northeastern United States." *Geographical Review* 72 (1982): 155–70.

Raitz, Karl B. "The Barns of Barren County." *Landscape* 22, no. 2 (1978): 19–26.

Sculle, Keith A., and H. Wayne Price. "The Traditional Barns of Hardin County,

Illinois: A Survey and Interpretation." *Material Culture* 25, no. 1 (1993): 1–27.

OTHER FARM STRUCTURES

Barkema, Alan, and Michael L. Cook. "The Changing U.S. Pork Industry: A Dilemma for Public Policy." *Economic Review* (Federal Reserve Bank of Kansas City) (2d quarter 1993): 49–65.

Eck, Paul. *The American Cranberry.* New Brunswick, N.J.: Rutgers University Press, 1990.

Gohlke, Frank, with a concluding essay by John C. Hudson. *Measure of Emptiness: Grain Elevators in the American Landscape.* Baltimore: Johns Hopkins University Press, 1992.

Gregor, Howard F. "Industrialized Drylot Dairying: An Overview." *Economic Geography* 39 (1963): 308–10.

Hart, John Fraser. "Change in the Corn Belt." *Geographical Review* 76 (1986): 51–72.

———. "Meat, Milk, and Manure Management in the West." *Geographical Review* 56 (1966): 118–19.

Hart, John Fraser, and Ennis L. Chestang. "Bright Tobacco: A Photo-Essay." *Focus* 41, no. 4 (1991): 1–7.

———. "Rural Revolution in East Carolina." *Geographical Review* 68 (1978): 435–58.

Hart, John Fraser, and Eugene Cotton Mather. "The Character of Tobacco Barns and Their Role in the Tobacco Economy of the United States." *Annals of the Association of American Geographers* 51 (1961): 274–93.

Hart, Samuel A. "Manure Management." *California Agriculture* (December 1964): 5–7.

Moran, Warren. "The Wine Appellation as Territory in France and in California." *Annals of the Association of American Geographers* 83 (1993): 694–717.

Noble, Allen G. "The Evolution of American Farm Silos." *Journal of Popular Culture* 1 (1980): 138–48.

Polczinski, Len C. "Ginseng (Panax quinquefolius L.) Culture in Marathon County, Wisconsin: Historical Growth, Distribution, and Soils Inventory." Master's thesis, University of Wisconsin at Stevens Point, 1982.

Roe, Keith E. *Corncribs in History, Folklife, and Architecture.* Ames: Iowa State University Press, 1988.

FARM SIZE AND FARM TENURE

Hart, John Fraser. "Change in the Corn Belt." *Geographical Review* 76 (1986): 51–72.

———. "Field Patterns in Indiana." *Geographical Review* 58 (1968): 450–71.

———. *The Land That Feeds Us*. New York: W. W. Norton, 1991.

———. "Nonfarm Farms." *Geographical Review* 82 (1992): 166–79.

———. "The Persistence of Family Farming Areas." *Journal of Geography* 86 (1987): 198–203.

Houston, J. M. "The Rural Landscape." In *A Social Geography of Europe*, 49–79. London: Duckworth, 1953.

Kollmorgen, Walter M., and George F. Jenks. "Suitcase Farming in Sully County, South Dakota." *Annals of the Association of American Geographers* 48 (1958): 27–40.

Smith, Everett G., Jr. "Fragmented Farms in the United States." *Annals of the Association of American Geographers* 65 (1975): 58–70.

SETTLEMENTS

Allix, Andre. "The Geography of Fairs: Illustrated by Old World Examples." *Geographical Review* 12 (1922): 532–69.

Berry, Brian J. L. *Geography of Market Centers and Retail Distribution*. Englewood Cliffs, N.J.: Prentice-Hall, 1967.

Dickinson, Robert E. "The Distribution and Function of the Smaller Urban Settlements of East Anglia." *Geography* 17 (1932): 19–31.

Evans, E. Estyn. *Irish Folkways*. London: Routledge and Kegan Paul, 1957.

Hart, John Fraser. "Small Towns and Manufacturing." *Geographical Review* 78 (1988): 272–87.

Hart, John Fraser, and Neil E. Salisbury. "Population Change in Middle Western Villages: A Statistical Approach." *Annals of the Association of American Geographers* 55 (1965): 140–60.

Hart, John Fraser, Neil E. Salisbury, and Everett G. Smith Jr. "The Dying Village and Some Notions about Urban Growth." *Economic Geography* 44 (1968): 343–49.

Hart, John Fraser, and Tanya Bendiksen. "Small Towns Can't Stop Growing." *CURA Reporter* (University of Minnesota) 19, no. 3 (1989): 1–5.

Hart, John Fraser, and Tanya Bendiksen Mayer. "Tough Times for Minnesota Small Towns." *CURA Reporter* (University of Minnesota) 21, no. 4 (1991): 8–11.

Meitzen, August. *Siedelung und Agrarwesen der Westgermanen und Ostgermanen, der Kelten, Römer, Finnen, und Slaven*. Berlin: Wilhelm Hertz, 1985.

Mikesell, Marvin W. "The Role of Tribal Markets in Morocco." *Geographical Review* 48 (1958): 494–511.

Spencer, J. E. "The Szechwan Village Fair." *Economic Geography* 16 (1940): 48–58.

THE LONG SHADOW OF THE CITY

Baerwald, Thomas J. "The Emergence of a New 'Downtown.'" *Geographical Review* 68 (1978): 308–18.

Campbell, John. "The Richest Crop." *Regional Review* (Federal Reserve Bank of Boston) 4, no. 3 (1994): 6–12.

Garreau, Joel. *Edge Cities: Life on the New Frontier.* New York: Doubleday, 1991.

Hart, John Fraser. "The Bypass Strip as an Ideal Landscape." *Geographical Review* 72 (1982): 218–23.

———. "The Modern Metropolitan Frontier." *Geography Research Forum* 13 (1993): 155–62.

———. "The Perimetropolitan Bow Wave." *Geographical Review* 81 (1991): 35–51.

———. "Redundant and Recycled Farmsteads." *Focus.* In press.

National Agricultural Lands Study. *Final Report.* Washington, D.C.: National Agricultural Lands Study, 1981.

———. *Where Have the Farmlands Gone?* Washington, D.C.: National Agricultural Lands Study, 1981.

Vesterby, Marlow, Ralph E. Heimlich, and Kenneth S. Krupa. *Urbanization of Rural Land in the United States.* Agricultural Economic Report 673. Washington, D.C.: U.S. Department of Agriculture, 1994.

RECREATION

Hart, John Fraser. "Population Change in the Upper Lake States." *Annals of the Association of American Geographers* 74 (1984): 221–43.

———. "Resort Areas in Wisconsin." *Geographical Review* 74 (1984): 192–217.

———. "A Rural Retreat for Northern Negroes." *Geographical Review* 50 (1960): 147–68.

Parsons, James J. "Slicing Up the Open Space: Subdivisions without Homes in Northern California." *Erdkunde* 26 (1972): 1–8.

Raitz, Karl B., ed. *A Tour of the Bluegrass Country.* Kentucky Study Series, no. 4. Lexington: University of Kentucky Department of Geography, 1971.

Tannenwald, Robert, ed. *Casino Development: How Would Casinos Affect New England's Economy?* Special Report no. 2. Boston: Federal Reserve Bank of Boston, 1995.

Tym, Alice Luthy, and James R. Anderson. "Thoroughbred Horse Farming in Florida." *Southeastern Geographer* 7 (1967): 50–61.

Index

ABOUT THE AUTHOR

John Fraser Hart was born in 1924 in Staunton, Virginia, and was raised in Virginia and Atlanta. He attended Hampden-Sydney College and the University of Georgia Extension Division in Atlanta, received an A.B. in classical languages from Emory University, studied geography at the University of Georgia, and completed an M.A. and a Ph.D. in geography at Northwestern University. His academic honors include the first Lifetime Achievement Award from the Southeastern Division of the Association of American Geographers in 1987, a John Simon Guggenheim Memorial Foundation Fellowship in 1982–83, the presidency of the Association of American Geographers in 1979–80, the editorship of the *Annals of the Association of American Geographers* from 1970 to 1975, an Award for the Teaching of Geography, College Level, from the National Council for Geographic Education in 1971, and a Citation for Meritorious Contributions to the Field of Geography from the Association of American Geographers in 1969. He is the author and editor of ten books, most recently *Our Changing Cities* (Johns Hopkins University Press, 1991) and *The Land That Feeds Us* (Norton, 1991), for which he received the Publication Award of 1992 from the Geographic Society of Chicago and the 1992 John Brinckerhoff Jackson Prize from the Association of American Geographers. John Fraser Hart is a professor of geography at the University of Minnesota.

Library of Congress Cataloging-in-Publication Data

Hart, John Fraser.
  The rural landscape / John Fraser Hart.
    p. cm.
  Includes bibliographical references and index.
  ISBN 0-8018-5717-1 (alk. paper)
  1. Rural geography.   2. Rural geography—United States.   3. Agricultural geography.
4. Agricultural geography—United States.   5. Landscape assessment.   6. Landscape assessment—United States.   7. United States—Geography.   I. Title.
GF127.H38   1998
333.76—dc21   97-28553
                                                                      CIP